NCRP REPORT NO. 57

INSTRUMENTATION AND MONITORING METHODS FOR RADIATION PROTECTION

Recommendations of the NATIONAL COUNCIL ON RADIATION PROTECTION AND MEASUREMENTS

Issued May 1, 1978

National Council on Radiation Protection and Measurements
7910 WOODMONT AVENUE / WASHINGTON, D.C. 20014

Library of Congress Catalog Card Number 78-62094
International Standard Book Number 0-913392-40-5

Preface

It was recognized many years ago that the original report on Radiological Monitoring Methods and Instruments—NCRP Report No. 10, published as National Bureau of Standards Handbook 51—needed to be rewritten. The task was assigned to the re-constituted Scientific Committee 7, which was the Committee responsible for the preparation of NCRP Report No. 10. This Committee found it difficult to keep ahead of the rapidly advancing technology of the 1960's and 1970's, and in their efforts to be complete and thorough, the members found completion of the text difficult. Another complicating factor was the appearance of government regulations, the impact of which had to be factored into the technique portion of the text.

This report supersedes NCRP Report No. 10. The text discusses, first of all, the fundamentals of survey and monitoring procedures for radiation protection purposes, then follows with the specifics of area survey methods. Personnel monitoring methods are described in detail in Section 4. There follows an extensive section on the different types of instruments used for the various types of surveys. Finally, there is a brief section on the principles of radiation accident monitoring. It is recognized that there are other texts devoted exclusively to this problem, but the NCRP is of the view that this text may well receive a greater circulation to facilities that may not yet possess the necessary expertise to control an unusual situation which may have the potential to cause unnecessary radiation exposure of individuals.

The Council has noted the action taken by the General Conference of Weights and Measures to make available special names for some of the units of the International System (SI) used in connection with ionizing radiation. Gray (Gy) has been adopted as a special name for the SI unit, joule per kilogram, for absorbed dose, absorbed dose index, kerma, and specific energy imparted. Becquerel (Bq) has been adopted as a special name for the SI unit of activity, (of a radionuclide). Since the transition from the special units currently employed—rad and curie—to the new special names is expected to take some time, the

Council has determined to continue, for the time being, the use of rad and curie. To convert from one set of units to the other the following relationships pertain:

$$1 \text{ rad} = 0.01 \text{ J kg}^{-1} = 0.01 \text{ Gy}$$
$$1 \text{ curie} = 3.7 \times 10^{10} \text{ s}^{-1} = 3.7 \times 10^{10} \text{ Bq (exactly)}$$

The present report was prepared by the Council's Scientific Committee 7 on Monitoring Methods and Instruments. Serving on the Committee during the preparation of this report were:

ARTHUR R. KEENE, *Chairman* (1960-1967)
EDWARD W. WEBSTER, *Chairman* (1968-)

MARGARETE EHRLICH
JOHN S. HANDLOSER
JACK S. KROHMER
RICHARD T. MOONEY
JACQUES OVADIA
WILLIAM C. ROESCH

Commander James A. Spahn of the NCRP's staff provided staff support for the Scientific Committee's efforts.

The Council wishes to express its appreciation to the members of the Committee for the time and effort devoted to the preparation of this report.

WARREN K. SINCLAIR
President, NCRP

Bethesda, Maryland
January 15, 1978

Contents

1. Introduction

The National Council on Radiation Protection and Measurements (NCRP) has made recommendations for the protection of individuals who may be exposed to radiation occupationally or otherwise. To implement the recommendations, it is necessary to examine the radiation environment of persons, to evaluate the factors determining the likelihood of their exposure, and to estimate the actual exposure. It is the object of this Report to describe methods and instrumentation currently available for these purposes.

The techniques, instruments, and practices described are applicable to all types of institutions concerned with radiation or radioactive materials. These institutions include industrial plants, scientific laboratories, universities, and hospitals or clinics. The radiation sources considered include x-ray machines, sealed and unsealed radioactive materials, low-energy accelerators, and low-power nuclear reactors. Additional and more elaborate instrumentation and measurement programs may be necessary and appropriate for specialized installations which may have unique problems beyond the scope of this Report.

This Report is intended for the guidance of all persons who are responsible for, or employed in, radiation protection programs. It is hoped that the Report will have value in the training of such persons. Much of this Report, particularly that part dealing with procedures for the conduct of a radiation survey, is directed toward the "surveyor". The "surveyor" may be defined as a person investigating the factors influencing the radiation exposure of individuals in or near installations in which radiation sources are present. The surveyor may inspect radiation protection features in a given installation, measure the radiation fields or contamination likely to produce radiation exposure, and evaluate actual or potential exposures in relation to permissible values. It should be recognized by the reader, and by employers and supervisors of radiation safety personnel, that competent performance of the inspections, measurements, and evaluations discussed in this Report requires personnel with varying degrees of training and experience depending upon the complexity of the particular job and the degree of excellence expected. Such persons include health physicists and health physics technicians, radiological physicists, and others.

In the simplest installations the necessary competence relates principally to the correct use of a survey meter, maintenance of records, and checking the operation of required safeguards. This level of competence will in some cases be possessed by the user or operator of small radiation sources who has specific training related to the safe operation of those sources. The highest level of competence requires wide experience of normal and abnormal conditions of radiation exposure in diverse installations, critical knowledge of the properties of radiation measuring instruments, and above all, the capacity for independent judgment on the acceptability of a given situation involving potential or actual exposure. Such competence is connoted by the term "qualified expert". This person should have at least the training and experience required by the American Board of Health Physics for certification in health physics, by the American Board of Industrial Hygiene in the radiological aspects of industrial hygiene, or by the American Board of Radiology for certification in radiological physics.

In the first section of this Report information is presented of a general character related to radiation surveys and instrumentation. The subsequent sections contain discussions of specific installations and types of measurement. Some procedures are used in surveying several types of installations, and some instruments are used in several types of measurements; for example, measurement of the radiation field and measurement of radioactive contamination. Hence, some information is presented in similar form under several headings. Such repetition has been held to a minimum, but is justified on the basis that the reader seeking guidance is likely to refer to a specific type of installation or a specific type of measurement.

There are three terms in this Report for which a special meaning is indicated by the use of italics:

shall is used to indicate that adherence to the recommendation is considered necessary to meet accepted standards of protection;

should is used to indicate a generally prudent practice to which exceptions may occasionally be made in appropriate circumstances;

exposure is used to indicate the quantity of ionization produced per unit mass of air by x or gamma radiation. The special unit of *exposure* is the roentgen.

2. Fundamentals of Survey and Monitoring Procedures

The selection and proper use of radiation detectors, and the development and application of effective monitoring methods, are of basic importance to any radiation protection program. Decisions are required on the need for monitoring the individual and for surveying the locations he may occupy, and on the methods to be used. The frequency of survey and monitoring, and the methods adopted, depend on many factors. The most important of these, discussed in this Section, are the type of radiation, the mode of radiation exposure, the level of radiation exposure relative to applicable permissible limits, and the extent to which reliance is placed on physical or procedural controls in achieving protection.

2.1 Protection Standards

The radiation protection standards operating in and around any given installation may derive from one or more of the following sources:

(1) the recommendations of national and international advisory bodies,
(2) enabling legislation based on these recommendations,
(3) regulations issued by governmental agencies, and
(4) administrative rules and controls imposed by those responsible for operating the installation.

2.1.1 *Recommendations of Advisory Bodies*

Dose limits for individuals occupationally exposed to radiation and for certain members of the general public have been recommended by the NCRP in Report No. 39 *Basic Radiation Protection Criteria* (NCRP, 1971b). These limits are expressed as "maximum permissible dose equivalents averaged over the designated organ or tissue" (MPD). They are established for the whole body and specific body organs. The dose equivalent is the product of absorbed dose and of factors chosen to allow for differences in the biological response to different types, energies, and spatial distributions of radiation. The principal such factor is the "quality factor" (Q) which depends on LET and thus on the kind of radiation or radionuclide involved. Another factor is the "distribution factor", which may be used to express modification of the biological effect due to non-uniform distribution of internally-deposited radionuclides.

NCRP Report No. 22 (NCRP, 1959) and addendum (NCRP, 1963) interpret the MPD in terms of the maximum permissible body burdens of radionuclides. The maximum permissible body burden (MPBB) is defined as that body burden of a radionuclide that, if maintained at a constant level, will produce the maximum permissible dose equivalent in the critical organ. The MPBB is dependent on the route of administration. It is also a function of time after a *single intake* of a radionuclide which will eventually produce the MPD in the critical organ. Report No. 22 also contains recommended values of the maximum permissible concentrations (MPC) of radionuclides in air and water for sustained intake. Intake at the level of the MPC, if continued for 50 years of adult life, would be expected to produce doses not in excess of the MPD.

Several other NCRP reports give specific guidance on radiation protection for various classes of equipment, such as x-ray machines, high energy electron accelerators, radionuclides in various forms, and neutron generators. These are referenced in Section 3.3 of this Report, which deals with specific installations.

Dose limits for several classes of exposed individuals have been recommended by the International Commission on Radiological Protection (ICRP, 1977). Recommendations of the ICRP on radiation protection for specific types of radiation sources are also available (e.g. ICRP, 1970).

Until 1971, the Federal Radiation Council[1] provided radiological guidance to all federal agencies in the United States and issued Radiation Protection Guides similar in meaning and magnitude to the MPDs recommended by the NCRP and the International Commission on Radiological Protection (ICRP).

2.1.2 *Governmental Regulations*

Federal, state, and local government agencies have issued regulations and codes that derive from enabling legislation, have the force of law, and may include penalties for failure to comply. Typical of these

[1] The Federal Radiation Council was abolished in 1971 and its function transferred to the Environmental Protection Agency.

regulations are those contained in Title 10, Code of Federal Regulations (CFR 1976d), which are applicable to persons licensed by the U.S. Nuclear Regulatory Commission. Other regulations apply to Department of Energy prime contractors. Many federal agencies, such as the Food and Drug Administration (Bureau of Radiological Health), the U.S. Postal Service, the U.S. Coast Guard, and the Department of Transportation, have issued rules and regulations pertaining to special aspects of radiation protection. A comprehensive set of regulations for several types of radiation sources is contained in *"Suggested State Regulations for Control of Radiation"* (BRH, 1974).

2.1.3 *Administrative Controls*

The management of installations in which radiation is employed usually imposes internal operating rules for establishing the dose-limiting standards and methods. These rules are usually derived from NCRP recommendations or current governmental regulations. Their objective is to assure compliance with the regulations and to provide for the elimination of unnecessary exposure. American National Standard N13.2 (ANSI, 1969a) provides brief guidance for administrative practices.

2.2 Dose-Limiting Methods

Methods for limiting human exposure in and near radiation installations can be classified as either physical (or engineered) safeguards or as procedural controls.

2.2.1 *Physical Safeguards*

Physical safeguards include all physical equipment used to restrict access of persons to radiation sources or to reduce the level of exposure in occupied areas. To minimize irradiation from external sources, protective devices include, but are not limited to, shields, barriers, locks, alarm signals, and source shutdown mechanisms. To minimize the irradiation from internally deposited radionuclides, typical protective devices are containers, hoods and glove boxes, ventilation, filtration and exhaust systems, protective clothing, and respiratory protective devices.

Planning and evaluation of physical safeguards *should* begin in the early phases of design and construction of an installation. Detailed inspection and evaluation of the radiation safety of equipment are mandatory at the time of the initial use of the installation. Additional investigations are necessary periodically to assure that the effectiveness of the safeguards has not decreased with time or as a result of equipment changes.

2.2.2 *Procedural Controls*

Procedural controls include all instructions to personnel regarding performance of their work in a specific manner for the purpose of limiting radiation exposure. Training programs for personnel are often necessary to promote observance of such instructions. Typical instructions concern mode of use of radiation sources, limitations on proximity to sources, exposure time, and occupancy of designated areas, use of protective clothing and devices, and the sequence or kinds of actions permitted during work with radiation sources.

Periodic area surveys and, in many instances, personnel monitoring are necessary to assure the adequacy of and compliance with established procedural controls.

2.3 Modes of Exposure

Irradiation of the body may result from radiation sources located outside the body (external exposure) or from radioactive materials deposited within the body (internal exposure) or simultaneous exposure from both. The equipment and methods used to assess dose to the body and its organs differ for these two modes of exposure.

2.3.1 *External Exposure*

As long as the radiation source remains external, exposure of the individual may be terminated by removing the individual from the radiation field, by removing the source, or by switching off a radiation-producing machine. If the external radiation field is localized, exposure to individuals may be limited readily by shielding or by denying access to the field of radiation or at least to the region of high radiation intensity.

2.3.2 *Internal Exposure*

Entry of radioactive materials into the body may occur by inhalation, by ingestion, or by absorption through the skin or through a break in the skin. Insoluble materials are usually excreted after ingestion, but after inhalation or entry through the skin, they may remain in the lungs or subcutaneous tissue or may migrate elsewhere in the body depending on the clinical form and particle size. Inhalation and ingestion frequently occur together since large insoluble particles in the tracheo-bronchial tree may be expelled by ciliary action and then swallowed. Soluble materials, after entering the body, are either excreted or deposited in one or more organs or tissues depending on their chemical and physical form and metabolic pathways in the body. Irradiation of the organs which the radionuclide traverses, or in which it is deposited, continues until the radioactive material is eliminated through translocation, excretion, or radioactive decay. Internal exposure can be avoided only by preventing the entry of radionuclides into the body and this is best achieved by strict confinement of radioactive materials at work or storage locations.

2.4 Evaluation of Radiation Measurement

An important index of the effectiveness of a radiation protection program is the comparison of the dose equivalent that is or may be received by individuals with the established maximum permissible levels. This comparison is usually needed for legal or administrative reasons. In addition, a program evaluation should include a judgment on whether the dose equivalent that is, or may be, received by individuals is as low as is reasonably achievable, and, if not, recommendations should be made concerning further protective measures (see paragraph 178, NCRP, 1971b).

As a general rule, it is not possible nor practical to measure dose equivalent (or even absorbed dose) at the point of interest in the human body. For external exposures, measurements are usually made of quantities such as *exposure*, kerma, or particle fluence from which absorbed dose, and subsequently dose equivalent, can be derived. Measurements of such quantities are made in free air (in the absence of the human body), in the vicinity of the human body, or on its surface, and subsequently converted to dose equivalent. The estimation of dose equivalent arising from internal emitters is particularly difficult. Usually internal exposure is assessed through estimation of body burden based on measurements of the activity concentrations in biological samples (bioassay), or on measurements of activity in the body by use of external counters. Alternatively, the potential body burden or the potential dose equivalent in body organs is assessed by comparing measurements of the concentration of radionuclides in air and water taken in by an individual with the maximum permissible concentrations that have been specified.

There are many sources of error and uncertainty in estimations of dose equivalent made on the basis of survey, monitoring, and bioassay measurements. These include the use of unsuitable measuring instruments, differences between the location of instruments and exposed individuals, insufficient knowledge of the exposure geometry, and, in the case of bioassay, uncertainty of the distribution of the activity in the body and of the time of exposure. Insufficient knowledge of the type of radiation or its energy spectrum may lead to incorrect choice of quality factor.

Dose equivalents derived by methods and assumptions described in this Report can be accepted as sufficiently accurate for radiation protection purposes if they do not exceed the applicable MPD. If calculated dose equivalents received, or expected to be received, are close to or greater than maximum permissible values, the circumstances of the measurements *should* be reviewed by a qualified expert and special consideration given to the conversion factors employed in deriving the dose equivalent.

3. Area Survey Methods

3.1 General

The area survey provides the basis for establishing the conditions under which occupancy of areas in the vicinity of radiation sources is permissible. The need for performing a radiation survey *should* be considered in any radiation installation where the radiation level may be greater than the prevailing background level. The specific circumstances in which a radiation survey is needed will vary with the times, conditions, use, and storage of the radiation sources. As a general guide, radiation surveys may be advisable in installations where:

(a) radioactive material with activity greater than 0.1 μCi and a radioactivity concentration greater than 0.002 μCi/g is present[2], or

(b) a device[3], in which charged particles are or may be accelerated through a potential difference greater than 10 kilovolts (kV)[4], is present.

The results of the survey may determine the conditions of source operation, occupancy limitations, shielding requirements, need for equipment to control contamination, and requirements for measurement of external and internal exposure received by individuals.

Area surveys *shall* be made:

(1) prior to the initial use of the radiation installation, or as soon as possible after radiation sources are brought into the area;

(2) whenever changes in procedures, equipment, shielding, or

[2] These activities are based on the maximum permissible body burden for the most hazardous radioelements, a surface dose rate of 2 mrad/h when beta rays of mean energy 0.5 MeV are emitted, and an exposure rate of 2 mR/h at 1 cm from a gamma-ray source.

[3] In certain widely-used devices operating at low kilovoltages, such as television receivers, a prototype survey to establish conformity with radiation emission limits is usually adequate. Standards for x-ray emission from electronic products, applicable before public use, have been issued by the Department of Health, Education and Welfare, Public Health Service, in the Code of Federal Regulations Title 21 Subchapter J under Public Law 90-602.

[4] 10 kilovolts is considered to be the minimum accelerating voltage for which x rays are likely to penetrate the source housing and deliver doses to the body that may produce harmful effects.

sources have occurred that may possibly cause an increase in exposure of personnel;
(3) periodically, to detect the effect of long-term changes in equipment, environment, and work habits;
(4) when an accident is suspected or after it has occurred.

The frequency of area surveys will depend upon the type of source, type of procedures, the workload, and personnel work habits, and *should* increase with the frequency of changes that could cause increased exposure.

Where radiation conditions are stable and predictable, a survey may be required as infrequently as once a year; where conditions are highly variable or unpredictable, where unsealed radioactive materials are handled directly, or where a radiation accident has occurred, daily area surveys or continuous area monitoring may be required. Continuous measurement of the radiation exposure rate of personnel is sometimes necessary; this process is known as *personnel exposure rate monitoring* and is discussed in Section 4.2.2.6.

3.2 Phases of Area Survey

A complete area survey usually includes the following phases:

(1) *Investigation.* Information is collected and examined regarding the radiation sources present, their intended use, the physical safeguards that are present or planned, and the applicable written procedural controls.
(2) *Inspection.* The surveyor personally verifies the presence of the sources and investigates their condition, their uses, the operability and integrity of the physical safeguards, and the adherence of personnel to established procedures.
(3) *Measurement.* The measurements may be of radiation fields, surface contamination, and/or airborne contamination depending on the nature of the installation.
(4) *Evaluation and Recommendations.* The results of measurements may be converted by means of operational factors into a form that can be directly compared with the applicable Maximum Permissible Dose Equivalents or Maximum Permissible Concentrations. The results of this comparison, together with the information obtained during the inspection, form the basis for an evaluation of the radiation safety status of the installation, and for recommendations regarding remedial action and resurvey after corrective action has been taken.

(5) *Records and Reports.* The results are recorded for future reference and reported to responsible persons.

3.2.1 *Investigations*

The first step in a complete area survey is an investigation to determine:

(1) The nature and purpose of the work performed in the installation.
(2) The type, number, and location of the radiation sources.
(3) The spatial relationship of the radiation sources to surrounding occupied areas.
(4) The type, location, characteristics, and purpose of physical safeguards and the extent upon which they are relied.
(5) The procedural controls and the extent upon which they are relied.

This information *should* be obtained by examining plans, drawings, charts, records, and written work instructions. An individual performing a survey within his own organization *should* make an investigation even though he believes he has firsthand knowledge of most of the facts to be gathered.

3.2.1.1 *Radiation Sources.* The following specific items of information on the radiation sources *should* be accumulated, as appropriate:

(1) The types and numbers of sources utilized (e.g., sealed sources, unsealed sources, or radiation generators), and their activity or rating.
(2) The types and energies of radiation produced, together with any modification of initial energy spectra by absorbers or moderators.
(3) The geometric size and position of radiation fields, and the direction of radiation beams relative to occupied spaces.
(4) The chemical composition and physical form of any radioactive materials.
(5) The potential for release or dispersion of radioactive material (including the possibility of induced radioactivity).

3.2.1.2 *Physical Safeguards.* The following specific information on the physical safeguards provided *should* be accumulated, as appropriate:

(1) The location, size, and construction of protective barriers and the level of radiation being transmitted through or scattered around such barriers.

(2) The nature of facilities for storage, handling, transportation, and disposal of radioactive sources.
(3) The design and construction of facilities for containment of unsealed radioactive materials, e.g., hoods or glove boxes.
(4) The location and design of the ventilation and exhaust systems, and of the associated filtration and radioactivity measuring equipment.
(5) The location and design of interlock, alarm, and emergency shutdown systems.
(6) The location and design of installed monitoring equipment.

3.2.1.3 *Procedural Controls.* The following specific information on the procedural controls provided *should* be accumulated as appropriate:

(1) The designation of areas as High Radiation Areas, Radiation Areas, or Controlled Areas.
(2) The degree of occupancy of positions inside and outside the installation during operation and the classification of persons exposed there as occupationally or non-occupationally exposed.
(3) The workload and use factor of each radiation source.
(4) The time an operator spends near each radioactive source.
(5) Operational procedures.
(6) The procedures for storage, handling, transportation, and disposal of radioactive sources.
(7) Any reports of previous area surveys.
(8) Any existing radiation protection rules and emergency plans of action.
(9) Types of instruction in safety procedures provided to inexperienced personnel.
(10) The identity of persons responsible for radiation protection.

3.2.2 *Inspection*

The inspection phase of an area survey is conducted to:

(1) Obtain a firsthand knowledge of the installation, personnel, equipment, surroundings, and practices relating to radiation protection.
(2) Determine where radiation measurements should be made.
(3) Determine the presence and effectiveness of each of the physical safeguards used for radiation protection.
(4) Determine the extent of compliance with procedural controls established for radiation protection.

Even though the individual making the survey may be familiar with the installation due to his employment or association, he *should* perform this inspection systematically.

3.2.2.1 *Radiation Sources.* The following specific items of information *should* be accumulated, as appropriate, during the inspection of radiation sources:

(1) The presence and location of each radiation source.
(2) The means provided for identifying each radiation source (i.e., serial number, rating, type, activity, size, shape, etc.).
(3) The physical integrity of the radiation source. For sealed sources, the methods, records, and results of leak testing *should* be evaluated.
(4) The apparent use made of each source.

3.2.2.2 *Physical Safeguards.* The following specific items of information *should* be accumulated, as appropriate, during the inspection of physical safeguards:

(1) The presence and likely adequacy of shielding or protective barriers.
(2) The possibility and consequences of inadvertent movement or removal of shields.
(3) The possibility and consequences of change in orientation or filtration of beams, or change in position of sources.
(4) The availability, condition, and mode of use of safety apparatus and special handling equipment; e.g., portable shields, remote control devices, hoods, dry boxes, protective clothing, showers.
(5) The possibility and consequences of improper air flow, spills, or introduction of radioactive materials into effluent from the installation.
(6) The adequacy of radioactive waste retention and/or disposal facilities.
(7) The suitability of the design of the installation for its use. This review may include flow of traffic, restriction of access or exit, ventilation, type of surface finish, location and type of water outlets, accessibility of shut-off valves or switches for air conditioning, electricity, water, gas, etc.
(8) The presence, correct functioning, and use of protective devices such as interlocks or visual or aural warnings, including signals indicating that the source is (or is about to be) turned on, evacuation alarms, ventilation failure alarms, and emergency off switches. The possibility of bypassing protective devices without adequate warning *should* be investigated.

3.2.2.3 *Procedural Controls.* The following specific items of information *should* be accumulated, as appropriate, during the inspection of procedural controls:

(1) The adequacy of labeling and posting of areas containing radiation sources, and labeling of containers of radioactive material.
(2) The adequacy of procedures for controlling personnel radiation exposure and spread of contamination during the handling, storage, transportation, and disposal of radioactive sources.
(3) The availability, appropriateness, correct functioning and use, and adequate provision for periodic recalibration of survey and monitoring equipment.
(4) The adequacy of routine survey and monitoring procedures.
(5) The existence, adequacy, and display of plans for action in anticipated emergencies, and the familiarity of personnel with these plans.
(6) The apparent level of training and alertness of personnel.

3.2.3 *Preparation for Measurements*

The choice of instruments and procedures to be used in making the necessary radiation measurements depends on the types of radiation or contamination to be measured and on the physical characteristics of the instruments. Guidance on selection of measurement instruments and equipment is provided in Section 5, and specific suggestions are made in connection with particular measurement problems in Sections 3.2.4, 3.2.5, and 3.3.

The individual making the required measurements *shall* establish whatever controls are necessary to insure that his own radiation exposure will be adequately measured and kept within permissible limits. In addition to the survey instrumentation required, he *shall* obtain and utilize personnel dosimeters, protective clothing, and respiratory protective equipment appropriate to the conditions that may be encountered. He *shall* ensure that radiation generators, source shielding mechanisms, or source handling equipment cannot be operated except under his control during the survey. He *shall* determine the general radiation condition prevailing in the area immediately upon entering and *shall* leave immediately or take other appropriate planned action if the radiation conditions change abruptly.

3.2.4 *Measurement of Radiation Fields*

In an area survey, measurements are made of radiation fields to provide a basis for estimating the dose equivalents that persons may receive. Changes in operating conditions (beam orientations, source outputs, etc.) can cause changes both in field intensity and pattern. Secondary radiations, discussed later, can cause field patterns to be very non-uniform. The number of measurements required depends on how much the radiation field varies in space and time, and how much people move about in the field. Measurements made at points of maximum intensity and at points of likely personnel occupancy under the different operating conditions are usually sufficient for estimates of dose equivalent to be made with accuracy adequate for protection purposes.

If the radiation pattern is fixed, as in many x-ray installations, few measurements are required. If, in addition, the potential dose equivalents are much less than the maximum permissible values, it suffices to determine the maximum dose equivalents that are likely to occur. On the other hand, if the radiation pattern is variable, such as during the removal of irradiated specimens from a reactor, many more measurements are required. In the extreme case, it may be necessary to monitor continuously while the work is in progress.

3.2.4.1 *Choice and Use of Instruments.* Radiation fields have components arising directly from the radiation source either in the useful beam or by leakage through the shield and barriers and they have additional components produced by scatter, induced activity, and other interaction phenomena. These secondary radiations (including scattered electrons, x rays, gamma rays, low-energy photons from Compton scatter, neutrons from photodisintegrations, etc.) may be different in type or energy from the incident radiation and may require the use of different instrumentation. The instrumentation used *shall* be capable of measuring any of these secondary radiations that are likely to be present in significant amounts.

Instruments used for the measurement of external exposure are normally calibrated in uniform fields of radiation delivered from a specific direction and having a particular energy spectrum. In most practical measurement situations, the energy, direction, or degree of uniformity of the field are different from those prevailing during calibration, and corrections may be required (see discussion in Section 5.1). The need for such corrections can, in some cases, be avoided by proper choice of the instrument.

The position(s) of maximum dose rate *should* be identified by scanning the areas with a small, sensitive detection instrument, such as a Geiger-Mueller or scintillation-type counter in the case of x-ray, gamma-ray, or beta-ray sources, or a BF_3 proportional counter enclosed in a moderator in the case of neutrons. Detection instruments are used to warn of the existence of radiation or radiation hazard and, as distinct from measuring instruments, usually indicate count rate rather than dose rate, *exposure* rate, or activity. Therefore, they *should* normally be used only to establish relative rates or activities. At positions of particular interest, individual determinations of dose or exposure rate *should* be made with measuring instruments. Dose integrating devices (dosimeters) may be mounted at points of interest and left for an extended period of time to improve the accuracy of the measurement.

Information concerning the dimensions, dose rate, and location of primary beams of radiation in relation to the source is important in the determination of direct external exposure from the beam and adequacy of protective measures. The dose or exposure rates within the beam at specific distances from the source *should* be measured and compared with expected values. Ionization chambers or thermoluminescent dosimeters are frequently used for these measurements. Techniques for determining cross-sectional area and position of x- and gamma-ray beams include use of radiographic films and fluorescent screens. Accelerator beams may be delineated by autoradiographic studies of induced radioactivity in foils.

Measurements close to radiation sources of small dimensions or of radiation transmitted through holes or cracks in shielding require special attention. The general location of defects in shielding *should* be determined by scanning with sensitive detection instruments. More precise delineation of the size and configuration of the defects can be obtained by using photographic film or fluorescent screens for x-ray, gamma-ray, or electron leakage. Measurements may then be made in any of three ways:

(1) An instrument may be used that has a detector volume small enough to ensure that the radiation field throughout the sensitive volume is substantially uniform.

(2) An instrument with a large sensitive volume may be used, if appropriate correction factors are applied. (See Section 5.1.1.2).

(3) Film may be used at the point of interest provided it has been properly calibrated for the types and energies of the radiations present.

3.2.4.2 *Conversion to Dose Equivalent Rate.* Survey instruments usually provide a measurement of the radiation incident upon a body in terms of *exposure* rate, particle flux density, or dose rate at the surface of the body. Before the dose equivalents that are expected for persons occupying the surveyed area are calculated (Section 3.2.6), the survey results *should* be translated into dose equivalent rates for one or more of the various critical organs of the body. This computation may require the application of physical and biological factors, depending on the units in which the initial measurements were made, the type and energy of radiation, and the critical organ of interest.

If the dose equivalent estimated for persons occupying the surveyed area is below the applicable permissible limits, the following simplifying assumptions can be made in many commonly encountered exposure situations involving only x-ray generators, or sealed or unsealed radioactive materials. These assumptions provide sufficient reliability since they are conservative.

(1) For gamma radiation and x rays, the *exposure* rate measured in R/h (whether made in free air, in the vicinity of the body, or on the body surface) may be taken to be numerically equal to the absorbed dose rate in rad/h and to the dose equivalent rate in rem/h, effective in all organs of the body.

(2) For beta radiation and electrons with energies less than 3 MeV, the absorbed dose rate in rad/h at the surface of the body may be taken to be numerically equal to the dose equivalent rate in rem/h, effective in the skin.

(3) For neutrons, a flux density of $4 \times 10^4 n/cm^2$ may be taken to be equivalent to an absorbed dose rate of 1 rad/h in all parts of the body, and to 10 rem/h dose equivalent rate for all body organs. The conversion from absorbed dose to dose equivalent for neutrons is discussed in more detail in Section 5.1.2.4.

When the estimated dose equivalents are close to, or in excess of, the MPD, the conversion of survey measurements to dose equivalent rate *should* be made more precisely by using physical and biological factors specifically applicable to the situation.

In complex situations (e.g., the vicinity of a high energy accelerator or a reactor) where photon, electron, neutron, and other radiations may all be present with a wide range of energies, the physical conversion factors used in the calculations will be highly dependent on the instruments used. Because of the wide variation in the magnitude of both physical and biological factors for different types and energies of radiations, the dose equivalent rate from each type *should* be computed separately. The dose equivalent rate contribution from the various radiation components should then be added for each organ of interest. In the case of neutron fields, however, sufficient accuracy may be provided by dose equivalent meters. For a discussion of these instruments see Section 5.1.2.4.

For conversion of these dose equivalent rates to expected dose equivalents, see Section 3.2.6.

3.2.5 *Contamination Measurements*

Electron accelerators operating at above 10 MeV and all heavy particle accelerators may induce activity in air or target materials exposed to the radiation beam. Surveys of accelerators, reactors, and installations using unsealed sources usually require frequent measurement of surface contamination levels and activity concentrations in air or water.

Contamination measurements are required around sealed sources only to determine whether they are leaking, and are not required in x-ray installations except where neutrons are produced.

3.2.5.1 *Surface Contamination.* Routine systematic searches *should* be conducted with sensitive detection instruments in any area where the possibility of surface contamination exists. All openings in source shields or contamination barriers and frequently handled items such as source manipulators, control switches, and knobs *should* be checked for contamination. Walls and horizontal surfaces such as floors, work benches, and shelves *should* be surveyed. After or during the search, count-rate measurements *should* be made at selected positions close to apparently contaminated surfaces.

The speed of search *should* be adjusted to the response time of the detection instrument and to the limits imposed by statistical variations. Rapid movement of an instrument with a long response time precludes meaningful readings. With audible (click or tone) or visible (flashing light) indication of count rate, the speed of search is limited only by statistical variations and by the ability to perceive small differences in signal frequency. For this reason, instruments used for contamination searches *should* include an aural or visual indicator.

Only surface contamination that is removable can enter the body; fixed contamination contributes only to external exposure. Therefore, the amount of removable surface contamination *should* be determined by a wipe test. The use of wipes is also recommended when a high background radiation level exists due to other sources in the vicinity.

Small pieces of paper, such as discs of filter paper, are rubbed over the surface, often with appropriate wetting or solvent material on the wiping surface, and examined at a remote location. Wipes can be examined for different types of radiation by using simple counting-room techniques, and can also be used to determine the radionuclides present by pulse-height analysis, absorption measurements, or radiochemical analysis as discussed in Section 5.2.3. The sensitivity of this type of survey is relatively high, but the accuracy of assaying the amount of radioactive material on the surface is poor. If wipes are used for surveying a particular operation or location, a standard method of sampling and assay *should* be established.

The individual making contamination measurements *should* be careful to avoid both exposure to himself and spread of contamination. If radiation detection instruments indicate that high levels of surface contamination may be present in an area, he *should* start wipe testing at the periphery of the area and proceed toward the point suspected of having high levels of contamination. He *should* wear adequate protective clothing during the survey, taking care to avoid contamination of hands, clothing, and radiation detection instruments. Instruments *should* be protected against contamination, by covering entirely with thin plastic material when only high energy beta or gamma radiation is present; the sensitive area of the detector *should* not be covered when alpha radiation is present or suspected. Shoe covers, gloves, and all instruments and equipment used in contaminated areas during an extensive survey *should* be checked for contamination routinely during the survey. Immediately after completion of the contamination survey, protective clothing *should* be removed and, together with the instruments and equipment used, checked for contamination.

Instrument readings and wipe test results *should* be recorded in the terms in which they were measured, together with a description of the methods and instruments employed. The contaminated areas *should* be clearly marked until the contamination has been removed, has decayed, or has been immobilized.

Additional guidance for measuring surface contamination may be found in NCRP Report No. 8 (NCRP, 1951a) and NCRP Report No. 30 (NCRP, 1964).

3.2.5.2 *Airborne Radioactivity.* For occupationally exposed persons, the most common mode of entry of radioactive material into the body is through inhalation. Since airborne radioactive material disperses rapidly and widely, prompt detection is important. In most cases, the investigative and inspection phases of the survey will show whether radioactive contamination of air is likely and what radionuclides may be present.

Some elements or compounds are normally gaseous; certain others are volatile or may produce volatile reaction products. All materials in dry, finely-divided form can become airborne and constitute a source of contamination. Solutions of non-volatile material may produce air contamination when heated, aspirated, aerated, or simply spilled. Long-continued low-level contamination inside a hood may eventually become airborne.

If occasional air contamination is likely in an installation, air sampling may be confined to those periods when release is possible; for example, when sources are being transferred between containers or heated. In areas where frequent or continuous air contamination is likely, the air *should* be sampled continually during periods of personnel occupancy.[5] The potential for nonoccupational exposure to airborne contamination *should* be assessed by sampling the gaseous exhaust stream from the facility. These measurements *should* be made downstream of any filters or other equipment provided to clean up the stream. The *actual* exposure of individuals cannot, in general, be deduced from the results of such sampling.

Determination of the concentration of radionuclides in air is made by measuring the radioactive material contained in known volumes of air. The air taken as a sample must be reasonably representative of the air inhaled. The sampling period *should* be sufficiently long to permit a useful comparison with permissible values. In the case of particulates, the sample is commonly obtained by filtration. Other techniques of particle collection include impaction and electrostatic precipitation. In the case of radioactive gases, the activity may be measured directly in the gaseous form, or after chemical or physical extraction. Alternatively, in the case of noble gases, a determination of the external exposure rate in a room will suffice. The measuring instrument may be an ionization chamber or counter sufficiently sensitive to, and calibrated for, the radiations from the particular radionuclides involved.

In counting filter paper from air samples, the counting rate may be unduly high initially because of the naturally occurring radon daughter products. Indeed, several methods of radon assay depend on filter counting. It is conventional to delay counting or to repeat the count 6 to 8 hours after taking the sample to allow for the decay of the radon daughter products. Correction for thoron daughters (^{212}Pb, $t_{1/2}$ 10.6 h) may be accomplished by a further repeat count after 24 hours (Healy, 1970). The count-rate from low-energy beta rays and alpha particles may be significantly reduced by absorption in the filter paper. Correction factors may be determined by measuring the activity in a sample after chemical or physical extraction. If continuous air samples are taken in a dusty atmosphere, the filter paper will become loaded and the air flow rate will decrease during the sampling process. Appropriate corrections *should* be made when the volumes of such air samples are calculated.

When an unknown mixture of radionuclides is present and the concentration is higher than the limits specified for unidentified nuclides, it may be necessary to identify the radionuclides [see Addendum 1 to NCRP Report No. 22 (NCRP, 1963)]. Commonly used methods

[5] In an alternative philosophy, the environment of personnel is controlled to levels well below the MPC and air sampling is done primarily to indicate when control is lost and remedial actions are needed.

are detector pulse-height analysis, analysis of the decay curve for the mixture, and measurements of the range of the emitted particles in an absorber.

In installations where air sampling is required, the survey *should* include an evaluation of the suitability of the sampling equipment used, the location of samplers, and the accuracy of the calibration of the sample-assay equipment. Where duct or exhaust stack sampling is employed, the adequacy of sampling lines or pipes *should* also be reviewed, particularly with respect to the need for isokinetic sampling. This will depend on the purpose of the sampling system and the distribution of radioactive particle sizes in the effluent. Information on sampling techniques is provided by ANSI (1969b) and on stack sampling by Sehmel (1967a, 1967b).

A discussion and description of air sampling equipment is included in Section 5.2.2. General guidance concerning techniques and procedures for measuring airborne contamination is given in NCRP Report No. 30 (NCRP, 1964) and NCRP Report No. 50 (NCRP, 1976c).

3.2.6 *Survey Evaluation and Remedial Recommendations*

3.2.6.1 *Evaluation of the Survey.* The measurements made in an area survey may be used directly to estimate the dose equivalent that would be received by personnel from external or internal exposure in the area during the time period in which the specific operating conditions prevail. For comparison with maximum permissible doses, it is frequently necessary to derive, from the short-term measurements, estimates of dose equivalent that would be received in extended periods of time such as a week or a month. This calculation requires the use of two time factors that

(a) express the time over which the source contributes to the dose, i.e., the use factor and,

(b) the probability of the area being occupied while the source is contributing to the dose, i.e., the occupancy factor.

In the case of external fields, the estimate obtained by the application of these factors is called the maximum expected dose equivalent (MED). For example[6], consider an area in which the dose equivalent

[6] In the case of medical x-ray installations, the MED may be calculated by employing terms described in NCRP Report No. 49 (NCRP, 1976b). For evaluation purposes a week is taken as the period of interest. In this case,

$$\text{MED} = \text{Dose Equivalent Rate} \times N \times U \times T,$$

where

rate is 32 mrem/h, the radiation field exists for 8 hours per week, the use factor (see NCRP Report No. 49) is ¼, and the location has an occupancy factor of 1/16 while the radiation field exists. The MED for a period of one week would be derived as follows:

$$\begin{aligned} \text{MED} &= 32\ (\text{mrem/h}) \times 8\ (\text{h/week}) \times 1/4 \times 1/16 \\ &= 4\ \text{mrem/week} \end{aligned}$$

The MED thus obtained *should* be compared with the MPD (for occupational or nonoccupational *exposure,* as appropriate) for the same period. After the derived MED is established, remedial action if necessary *should* be recommended in accordance with the principle stated in Section 3.2.6.2.

The average airborne or waterborne radioactivity concentration to which people might be exposed over an extended period *should* be determined for locations of interest by applying the operational modifying factors to the concentrations in air and water measured during the survey. For example, consider a location where the measured airborne concentration is $1 \times 10^{-9}\ \mu\text{Ci/cm}^3$ for 10 hours in each 40-hour week and an occupationally exposed individual occupies this location half-time during those 10 hours. Then the average concentration to which the individual is exposed is calculated as

$$1 \times 10^{-9}\ \mu\text{Ci/cm}^3 \times \frac{10}{40} \times \frac{1}{2} = 0.125 \times 10^{-9}\ \mu\text{Ci/cm}^3$$

For the nonoccupational situation, similar calculations can be made by using appropriate time and occupancy factors.

For occupational *exposure,* the expected average concentration of identified or unidentified radionuclides in air or in water *should* be compared with the 40 h/week Maximum Permissible Concentrations $(\text{MPC})_A$, $(\text{MPC})_W$, $(\text{MPCU})_A$, and $(\text{MPCU})_W$ appearing in NCRP

$$N = \text{“on time”}\left(\frac{\text{min}}{\text{week}}\right) = \frac{\text{workload}\left(\frac{\text{mA. min}}{\text{week}}\right)}{\text{current (mA)}}$$

U = use factor (dimensionless)

T = occupancy factor (dimensionless)

Current = current used during survey measurement (mA)

Dose Equivalent Rate-rem/min (derived from survey measurement)

The unit of MED is rem/week. Report No. 49 provides typical workload, use, and occupancy factors for various kinds of installations, satisfactory for use when actual values are not available.

Report No. 22 (NCRP, 1959) and its addendum (NCRP, 1963). For nonoccupational exposure, the MPC values to be used *shall* be one-tenth of the 40 h/week values in NCRP Report No. 22 when maximum occupancy is 40 h/week, and one-tenth of the 168 h/week values when maximum occupancy exceeds 40 h/week.

Surface contamination *should* be evaluated on the basis of guidance appearing in NCRP Report No. 30 (NCRP, 1964), e.g., Table 6, "Suggested Levels of Significant Contamination", to determine the necessity for decontamination.

3.2.6.2 *Recommendations for Remedial Action.* The evaluation of survey results may indicate deficiencies in the radiation protection program. Recommendations for remedial action *should* be made to correct deficiencies so as to maintain dose equivalents at levels which are as low as reasonably achievable (Paragraph 178, NCRP, 1971b; ICRP, 1973). Recommendations may be directed toward changes in

(1) operational factors, particularly occupancy time, equipment use time, or mode of use;
(2) shielding, particularly size, thickness, type of material, or location;
(3) manipulative equipment, particularly relating to speed of operation and distance of personnel from sources;
(4) procedural controls, particularly those that eliminate unnecessary personnel exposure or contamination;
(5) personnel protection or warning devices;
(6) survey and monitoring procedures;
(7) personnel monitoring equipment and survey equipment;
(8) plans of action for accidents or emergencies.

The need for resurvey following remedial action *shall* be determined. Ordinarily, a resurvey *should* be recommended if the remedial action concerns changes in physical safeguards. A resurvey may not be necessary if the remedial action concerns only procedural controls.

In some situations an area survey is performed immediately prior to, or during the performance of, specific work, a procedure requiring that the surveyor evaluate results during the survey if remedial action is to be timely and effective. Dose equivalent rates may then be compared with maximum short-term rates (e.g., in mrem per day) derived from the MPD for external exposure. This comparison is used to establish the maximum permissible work time of the individuals and to assure that a substantial fraction of the MPD is not received in a short time.

The results of measurements of airborne and surface contamination must often be evaluated immediately. The evaluation may suggest desirable changes in protective equipment or operational factors.

3.2.7 *Record of the Survey and Report*

Radiation survey records are needed to assess the effectiveness of the radiation protection program and may be required to aid in the interpretation of the results of personnel monitoring. Written records *shall* be maintained of all mandatory area surveys, including the results of inspections, whether made by individuals from outside the organization concerned or by individuals from within the organization. These records *should* contain sufficient detail to be meaningful after the passage of many years. However, to minimize the detail recorded in a given survey, separate records may be maintained for description of installations, instruments, methods, and personnel, where these are constant for a group of surveys. Nevertheless, they *should* be referenced in the report. When applicable, the following information *should* be included in the records:

(1) The location, identification, and function of the installations surveyed;
(2) the identity of the person responsible for radiation safety in the installation;
(3) a description of the radioactive material, radiation generating machines, or reactors used;
(4) the time and date of the survey;
(5) the specific physical safeguards and procedural controls inspected;
(6) the type and energy of the radiations involved;
(7) the mode of use of the source(s) during the measurements;
(8) the methods and instruments used in measurements; for contamination surveys, the method of obtaining the wipe or air samples; for waste surveys, the nature of the gas or liquid samples and the methods of obtaining them;
(9) the location at which measurements or wipes were made, or where liquid and air samples were obtained, either by written description or by sketches;
(10) the results of the measurements with the quantities and in the units in which they were obtained (e.g., mR/h, n cm^{-2} s^{-1}, mrad/h, mrem/h, or count/min). For contamination surveys the units are count/min per area wiped or count/min on the wipe if the area is unknown or unimportant; in waste surveys, the units are count/min per unit volume of air or liquid. Appropriately calibrated instruments permit the recording of results in units of activity instead of count/min;
(11) dose or activity calculations including calibration factors and the assumptions made;

(12) dose equivalents or dose equivalent rates where needed, and activity concentrations or levels in occupied areas;
(13) comments on the accuracy and reliability of results, if required;
(14) the identity of the persons performing the survey;
(15) the evaluation of leak test records, particularly test frequency and source contamination levels;
(16) the evaluation of the survey measurements, calculations, and inspections, including the comparison with appropriate MPDs and MPCs, and recommended contamination limits;
(17) a statement of conclusions from the survey and of whether the findings are in compliance with present radiation safety rules and regulations and with recommended good practice;
(18) recommendations regarding remedial actions and resurvey.

Records of surveys *should* be kept for a time and in a manner that assures compliance with applicable government regulations. The minimum storage period *should* be five years.

In some circumstances, particularly those where the survey results indicate that remedial action is required or where compliance with government regulations must be demonstrated, a written report summarizing the findings of the survey *should* be given to the person responsible for operating the installation. This summary *should* include the following information obtained from the survey record:

(19) An identification of the installation and locations surveyed;
(20) the dose equivalents or dose equivalent rates and activity concentrations at those locations and the corresponding mode of source use;
(21) the evaluation of the survey, including the comparison of the results with applicable limits;
(22) recommendations for remedial action, and need for resurvey.

3.3 Recommendations for Specific Installations

Section 3.2 presents recommended survey methods applicable to radiation installations in general. This section presents recommendations unique to several specific types of radiation installations. These recommendations supplement but do not substitute for the methods set out in Section 3.2.

3.3.1 *Medical X-Ray Installations*

Radiation protection is achieved in most medical x-ray installations by application of the physical safeguards and procedural controls recommended by the NCRP in Reports 33 and 49 (NCRP, 1968a; 1976b) and promulgated in federal standards for medical x-ray equipment (CFR, 1976b)[7]. Surveys *should* be performed periodically to assure that these recommendations are met.

The following recommendations are concerned primarily with the protective aspects of installations rather than the precision of technical performance. Tests for compliance with federal performance standards are published by the Bureau of Radiological Health (DHEW, 1974).

Measurements made during an area survey are dependent on the *exposure* produced by the x-ray tube when operated at specific kilovoltage, tube current, and timer setting. Measurement of the *exposure* rates in the useful beam *shall* be made, at a specified distance from the tube target, for several kilovoltages and current settings routinely used. The surveyor *should* evaluate the measured *exposure* rates on the basis of expected values.

Measurements of leakage radiation through the tube housing are usually unnecessary for modern x-ray tubes since the housings are generally designed to conform to the specifications of diagnostic-type or therapeutic-type protective housings. If the housing specification is in doubt, leakage measurements *should* be made. Recommended methods are contained in NCRP Report No. 33 (Sections 3.2.2 and 3.4.2). During the initial survey, qualitative tests for inadvertent gaps in the shielding of x-ray tube housings *should* be made, since tube housings with incomplete shields have, on rare occasions, been installed.

Measurements of x-ray transmission through primary and secondary barriers *shall* be made with those beam orientations and field sizes used in practice that will result in the greatest *exposure* rate at the point of measurement. Measurements in the direction of the useful beam *shall* be made without a phantom in the beam. Measurements of scattered radiation *shall* be made with a phantom in the useful beam simulating the object being irradiated.

3.3.1.1 *Medical Radiography Installations.* Medical radiographic equipment is generally installed in a shielded room with its control located either outside the room or in a shielded booth within the room. Peak tube potentials range from 25 to 150 kV. Tube currents may be as high as 2000 mA. Exposure times may be as low as one millisecond and rarely exceed a few seconds. The workload in milliampere-minutes per week can range from less than 10 in a very small office to perhaps 1000 in a busy hospital radiographic room. Although many different orientations of the useful beam are possible, in most installations it is directed only at the floor or certain wall areas. Probable use factors for

[7] The federal standards for diagnostic x-ray systems are applicable to equipment manufactured after August 1, 1974.

these directions *shall* be estimated, when it is likely that the MED will approach or exceed the MPD (see Section 3.2.6.1).

The surveyor *shall*, in general, determine whether the installation complies with the recommendations of NCRP Report No. 33 (NCRP, 1968a) and with those protective provisions of 21 CFR 1020.31 susceptible to field testing. In particular he *shall* determine that:

(1) The radiographic field sizes in use equal, or can be adjusted to equal, the sizes of the films[8] employed, within 2 percent of the target-film distance used. The presence and use of suitable diaphragms or cones or a calibrated adjustable collimator *shall* be ascertained and field coverage by the film at the film distance determined. In installations with automatic collimation, the correct operation of the device *shall* be tested (21 CFR 1020.31). Field coverage may be determined by calculations or by exposure of film[9], a fluorescent screen, or fluorescent strips. Some installations have adopted the practice of limiting fields so that all the edges of the field are included on the film; proper field coverage can then be verified by inspection of recently exposed films.

(2) For collimators in which visual definition of the field is provided, the misalignment between the edges of the visual field and the x-ray field is less than 2 percent of the target-film distance. The extent of misalignment may be measured on a radiograph in which the edges of the visual field have been marked prior to exposure by radio-opaque objects placed inside the visual field.

(3) The axis of the x-ray beam when perpendicular to the film is aligned with the center of the film within 2 percent of the target-film distance. This test *should* be performed, when feasible, with the film centered in the cassette-holder (Bucky tray) of the radiographic table and/or in the vertical cassette-holder.

(4) Measures are taken to protect the gonads of patients from exposure to the useful beam in those examinations where such exposure is not clinically necessary (CFR 1976e).

(5) The total tube filtration complies with the values recommended in NCRP Report No. 33. This compliance may be determined by demonstrating that the half-value layer equals or exceeds values stated therein.

(6) The duration of x-ray beam exposures is in agreement with

[8] Film is used generally in this section to exemplify radiographic image receptors.

[9] Rapid and economical field size measurements can be obtained with direct-print photographic paper and the use of fluorescent light or direct sunlight for development (DHEW, 1971).

the setting of the timer. Agreement may be demonstrated by means of a radio-opaque spinning disc containing a small peripheral hole which is placed in the x-ray field during radiography; or by means of *exposure* measurements with constant tube current in the useful beam that show proportionality between *exposure* and nominal time for a series of timer settings.

(7) At any of the current settings in normal use, the *exposure* per milliampere-second (mR/mAs) does not differ by more than 20 percent from the mean value of mR/mAs for the range of currents used. The measurements *should* be made at a normally-used tube kilovoltage and at a defined distance in air from the tube target.

(8) The exposure switch can be conveniently operated only when the operator is in a shielded position; or, with mobile equipment, the operator can stand at least 6 feet from the patient, the tube, and the useful beam when making exposures.

(9) Positive indication of the production of x rays is provided and can be readily discerned by the operator.

(10) The operator can see and communicate with the patient during exposure.

(11) The target-skin distance is limited to not less than 30 cm (12 inches) on mobile equipment.

For measurements of stray radiation, the x-ray tube *should* be operated at the maximum kVp setting in routine use and at a current low enough to avoid overloading the tube. The thermal characteristics of the x-ray tube *should* be considered, and reference made to the tube manufacturer's rating and cooling charts. The chosen current *should* permit sufficient exposure time to establish the *exposure* rate at any given location and a sufficient number of measurements to determine the radiation field. Frequently, the controls may not permit operation with currents lower than 50 to 100 mA, and therefore measurements must be made with short exposure times and high *exposure* rates. Such measurements are best performed with a high-range rate instrument (up to 10 R/h) with a short time-constant, preferably not exceeding one second or 30 percent of the exposure time, whichever is shorter. Ionization chamber rate meters are the most suitable instruments. Counting instruments should not be used because they may block when exposed beyond their range and because their response is often too energy-dependent for this purpose. Adequate energy independence can be obtained with condenser chambers and rate meters operated in the integral mode, but the sensitivity of many available chambers is limited by their volumes.

The survey *shall* include an evaluation of the personnel monitoring procedures. For most radiographic procedures it is not necessary to remain in the room during exposures, and as a result personnel dosimeters may not be required. However, during certain special procedures, such as urological radiography and angiography, and during the operation of mobile units, personnel may have to remain in the room. Under these conditions personnel dosimeters or other monitoring devices are mandatory even though protective aprons and gloves are worn.

3.3.1.2 *Dental Radiography Installations.* In most dental offices the x-ray equipment is installed near the dental chair. Peak tube potentials range from 40 to 100 kV, with tube currents from 5 to 15 mA. Many different beam orientations are used in practice. A protective booth for the operator is usually not required, but a protective barrier may be needed when the workload is high and the distance between operator and x-ray tube is short. Representative workloads range from 10 to 80 milliampere-minutes per week. With these workloads, masonry walls provide adequate room shielding. However, the present trend is to use interior drywall partitions (gypsum board) that provide little attenuation. Recommendations for radiation protection in dental installations may be found in NCRP Report No. 35 (NCRP, 1970) and in (CFR 1976b).

The surveyor *shall* determine that:

(1) The distance from the target to the end of the cone is at least 18 cm (7 inches) if the operating voltage is above 50 kVp, and at least 10 cm (4 inches) if 50 kVp or less.

(2) The field size at the end of the cone is restricted to 7 cm (2–¾ inches) or less in diameter for a source-skin distance of 18 cm (7 inches) or more, and to 6 cm (2 inches) or less for a source-skin distance less than 18 cm (7 inches). Photographic film or a long-lag fluorescent screen is useful for this purpose.

(3) The total tube filtration complies with the values recommended in NCRP Report No. 35. This may be determined by demonstrating that the half-value layer at a given tube voltage is not less than values stated therein.

(4) The operator can stand at least 2 m (6 feet) from the patient and well away from the useful beam while making exposures, or entirely behind an adequate shield.

(5) The exposure is terminated at a preset time and the timer provides a reproducible exposure time to at least 1/60th second. A radio-opaque spinning disc is recommended for determination of timer accuracy.

In addition, the exposure at the distal end of the cone *shall* be measured for a typical set of technique factors used in the installation. Condenser ionization chambers are recommended for this purpose.

The surveyor *should* evaluate the measured exposure on the basis of expected levels.

For stray radiation measurements, the x-ray tube *should* be operated at the kVp and mA settings used routinely. Long-term measurements at or beyond the room walls and doors may be made by posting film dosimeters or other integrating dosimeters in those locations for extended periods.

3.3.1.3 *Medical Fluoroscopy Installations.* Fluoroscopic equipment is usually operated at tube peak potentials of 60 to 125 kV with tube currents of up to 5 mA. During conventional fluoroscopic procedures it is necessary for personnel to remain in the room. Radiation scattered from the patient and various parts of the equipment may result in relatively high exposure to operating personnel, even though the primary barrier in the viewing mechanism attenuates the useful beam over its entire cross-section.

The surveyor *shall* in general determine whether the installation complies with the recommendations of NCRP Report No. 33 (NCRP, 1968a) and with those provisions of 21 CFR 1020.32[7] (CFR, 1976b) susceptible to field testing. In particular he *shall* determine that:

(1) The entire cross section of the useful beam is intercepted by the primary protective barrier of the fluoroscopic image receptor at any distance of the receptor from the x-ray source. A small piece of fluorescent screen mounted on a long handle and used in a darkened room is useful for detecting the presence of the unattenuated useful beam beyond the limits of the primary barrier [see item (14) for necessary transmission measurements].

(2) The exposure is automatically terminated (prevented) when the primary barrier is removed from the useful beam.

(3) In fluoroscopes used only for fluoroscopy or recording of images through an image intensifier, the dimensions of the x-ray beam cannot exceed the dimensions of the visible area of the image receptor at any distance of the receptor from the x-ray source. The difference between beam dimension and receptor dimension may be estimated by radiographically recording the maximum beam size in a plane close to that of the receptor and measuring the maximum object dimension observable fluoroscopically in the same plane. Limitations are placed on the difference by 21 CFR 1020.32 (CFR, 1976b).

(4) The center of the x-ray beam is aligned with the center of the visible area of the image receptor.

(5) In fluoroscopes equipped with spot-film devices subject to 21

[7] The federal standards for diagnostic x-ray systems are applicable to equipment manufactured after August 1, 1974.

CFR 1020.31, the automatic field size adjustment operates correctly and meets the requirements for alignment between the collimated useful beam and the edges of the selected portion of the film.

(6) The source-skin distance on stationary fluoroscopes is not less than 30 cm (12 inches) or is not less than 38 cm (15 inches) on fluoroscopes subject to 21 CFR 1020.32 (CFR 1976b). If an object of known size is placed at the skin position and its image is recorded at a known distance beyond the object, the source-skin distance may be calculated from the magnification factor of the image.

(7) The source-skin distance on mobile fluoroscopes is not less than 30 cm (12 inches).

(8) A cumulative timing device is used to signal the passage of a predetermined time period or to interrupt the operation of the apparatus after a predetermined time period, such period not exceeding 5 minutes.

(9) The total tube filtration complies with the values recommended in NCRP Report No. 33. This may be determined by demonstrating that the half-value layer is not less than values stated therein.

(10) A cover for the film slot immediately under the table and a shield attached to the viewing device or the table to intercept scatter are provided.

(11) Light exclusion from the room and provision for dark-adaptation are adequate when a non-image-intensified fluoroscope is in use.

(12) Auxiliary protective devices (aprons, gloves, and shielded chairs) are free from defects in the shielding provided. Discontinuities of the shielding in such devices can be detected readily by fluoroscopic or radiographic examinations, and sometimes by visual or manual examination.

For radiation measurements, the x-ray tube *shall* be operated at those kVp and mA settings used routinely and also at the maximum possible settings. Measurements *shall* be made of:

(13) The *exposure* rate in the center of the useful beam at the table top or panel. Condenser ionization chambers are recommended for this purpose.

(14) In all fluoroscopic systems the *exposure* rate at 10 cm from any accessible surface of the fluoroscopic imaging assembly beyond the plane of the image receptor, and at commonly occupied areas beside the fluoroscope. An ionization chamber rate meter is recommended for these measurements.

Measurements under (14) *should* be made with a unit density phantom about 12 inches (30 cm) × 12 inches (30 cm) × 8 inches (20 cm) thick in the beam (or an equivalent phantom)[10] and with the maximum possible field size.

If the system contains an *automatic brightness control* (frequently included in television and cinefluorographic systems), the measurements under (13) *shall* be made: (a) with a phantom as described above in the useful beam, and (b) with a lead plate at least 0.5 mm thick completely intercepting the useful beam. Measurements under condition (b) are necessary to obtain the maximum possible *exposure* rate. The measurements *should* be made with a small field size (e.g., 5 cm × 5 cm at the table-top or panel) in order to minimize the production of backscatter.

3.3.1.4 *Therapy Installations.* X-ray tube potentials in therapy installations range from peaks of 10 kV to many thousand kV with currents in the range of 0.25 to 30 mA. The treatment room usually requires structural shielding at peak potentials above 60 kV.

In all therapy installations, the surveyor *shall* determine the appropriate use factor for each primary barrier if this is a significant factor in evaluating its adequacy, and *shall* investigate the need for and adequacy of restriction on the beam orientation such as limit switches. The surveyor *shall* also determine that:

(1) The radiation transmitted by the collimation system and the filter slot conforms with the recommendations of NCRP Report No. 33 (NCRP, 1968a).

(2) Provisions have been made to minimize the possibility of error in filter selection.

(3) A device is provided that terminates the exposure at a preset value or after a preset time, regardless of other methods of exposure termination.

(4) Provisions are made for observing and communicating with the patient from the control area during treatment.

(5) The production of radiation is shown by an indicator on the control panel.

(6) The x-ray machine is provided with a locking device to prevent unauthorized use.

(7) The radiation "ON" switch opens when the power is disconnected.

In addition, for equipment capable of operating at tube peak potentials above 150 kV, the surveyor *shall* determine that:

[10] To check compliance of the exposure rate beyond the primary protective barrier with the provisions of 21 CFR 1020.32, the phantom *shall* be made from type 1100 aluminum and with dimensions 20 cm × 20 cm × 3.8 cm.

(8) The control panel is located outside the treatment room or in an enclosed protective booth.

(9) All access doors to the treatment area are provided with interlocks that operate in accordance with the recommendations of NCRP Report No. 33 (Section 3.4.1(m)) and NCRP Report No. 49 (Section 6.1.5).

In addition, for equipment capable of operating at a tube peak potential above 500 kV, the surveyor *shall* determine that:

(10) The production of radiation is shown by an indicator in the treatment room.

(11) Emergency "OFF" switches are located at appropriate positions in the treatment room.

For Grenz-ray treatment (tube peak potential below 20 kV) and contact therapy (tube peak potential 40 to 60 kV), structural shielding is usually not required and the operator often is in close proximity to the patient during treatment. The surveyor *shall* determine that:

(12) A protective apron is available for the operator during treatments.

(13) Contact therapy tubes are not hand-held, or, if hand-held, protective gloves are available.

For measurements of stray radiation in these installations, the use of thin-walled ionization chamber rate meters is recommended.

Maximum possible exposure rates in therapy installations *shall* be measured with the machine operated at the maximum kVp setting and at the maximum current setting for that kVp and with the minimum filter used in practice. In addition, typical *exposure* rates for the more commonly used machine settings *should* also be determined.

3.3.2 *Non-Medical X-Ray Installations*

Recommendations for radiation protection in x-ray installations for non-medical purposes are presented in NBS Handbook 114 (ANSI, 1975). The classification therein of non-medical x-ray installations into protective, enclosed, unattended, and open installations sets forth minimum standards of protection.[11] With enclosed and protective installations, protection is accomplished primarily by physical safeguards including a shielding enclosure. Unattended installations consist of automatic, single-purpose devices that operate without person-

[11] Some protective installations manufactured on or after 10 April 1975 are subject to federal performance standards for cabinet x-ray systems contained in 21 CFR 1020.40, particularly regarding ports, apertures, safety interlocks, controls, and indicators. These standards are also applicable to x-ray systems manufactured on or after 25 April 1974 for the inspection of carry-on baggage (CFR, 1976c).

nel in attendance, usually in uncontrolled areas. Open installations require stringent procedural controls that include a conspicuously posted perimeter since extensive physical safeguards are not possible due to operational requirements. Surveys *should* verify that procedural controls and physical safeguards required by the category of the installation are present and effective.

3.3.2.1 *Protective and Enclosed Installations.* In all protective and enclosed installations, the surveyor *shall* determine that:

(1) The equipment enclosure is shielded such that, in the case of the protective installation, the *exposure* rate at any accessible region 5 cm from the outside surface of the enclosure cannot exceed 0.5 mR in any one hour; and in the case of the enclosed installation, the *exposure* in occupied regions 30 cm from the outside surface of the enclosure does not exceed 10 mR in any one hour and in normally unoccupied regions 30 cm from the outside surface, does not exceed 100 mR in any one hour.

(2) Interlocks of fail-safe design are provided to prevent access to the enclosure during irradiation.

(3) Methods exist for assuring that the enclosure is unoccupied before irradiation including, where needed, the provision of fail-safe audible or visible warning signals within the enclosure (ANSI, 1975).

(4) Warning signals, a suitable exit, and a labeled means of preventing or quickly interrupting irradiation are provided in the enclosure.

In enclosed and protective installations, the surveyor *shall* also estimate the use factor for all protective barriers.

3.3.2.2 *Unattended Installations.* For unattended installations, the surveyor *shall* ascertain that:

(1) The source enclosure is shielded such that the *exposure* at any accessible region 30 cm from the outside surface of the enclosure cannot exceed 2 mR in any one hour under normal operating conditions; or the *exposure* rate conforms to overriding governmental or local regulations.

(2) Occupancy near the device is restricted so that *exposure* to an individual cannot exceed 0.5 R in one year.

(3) The device carries a sign warning of the presence of x rays or radioactive material.

(4) Access to areas with *exposure* levels exceeding those specified in item (1) is provided only through the use of restricted keys or tools.

(5) The open and closed positions of any shutter for radiation attenuation are clearly indicated.

(6) A visible signal indicates when x rays are produced.

3.3.2.3 *Open Installations.* In open installations, the surveyor *shall* determine that:

(1) The equipment is established within a perimeter that limits the area in which the *exposure* can exceed 100 mR in any one hour, and which is conspicuously posted with "High Radiation Area" signs.
(2) The procedures for denying access to the posted area during irradiation are adequate and enforced.
(3) Operators are instructed in the safe operation of the installations.

For enclosed and open installations, and for protective installations where personnel can enter the exposure area, the surveyor *shall* determine whether all workers involved in the use of radiation apparatus in the installation carry personnel monitoring devices and whether other personnel need to be monitored.

3.3.2.4 *Analytical X-Ray Equipment.* This class of equipment includes x-ray diffraction and fluorescence analysis devices in which the radiation source is either an x-ray tube or a radionuclide emitting low-energy photons. Typical x-ray tube potentials are 25 to 50 kVp in diffraction equipment and 25 to 100 kVp in fluorescence analysis equipment. Tube currents are usually of the order of 20 mA. Several ports with very low inherent filtration are normally utilized. The main protection problem is inadvertent exposure to the useful beam. X-ray *exposure* rates of 4×10^5 R/min at a port have been reported for diffraction tubes (Howley and Robbins, 1967). An accidental exposure of a few seconds to the useful beam may cause severe and permanent injury to the part of the body exposed. Discussions of radiation safety in these units are presented by Lindell (1968) and Moore *et al.* (1971).

Surveys of analytical x-ray equipment *shall* be conducted to determine conformity with the safety standards recommended in NBS Handbook 111 (ANSI, 1972b). The standard recognizes two classes of equipment: Enclosed Beam X-Ray Systems in which all possible x-ray paths are fully enclosed, and Open Beam X-Ray Systems in which one or more x-ray path is not fully enclosed. The evaluation of radiation surveys will be different in these two classes.

For all equipment, the surveyor *shall* ascertain that:

(1) Beam traps or other primary beam shields are in place and the transmitted *exposure* rate cannot exceed 0.25 mR/h under normal operating conditions.
(2) The *exposure* rate in accessible regions 5 cm from the surface of enclosures containing accessory equipment such as high voltage rectifiers cannot exceed 0.25 mR/h.
(3) A readily visible fail-safe indicator that x rays are being pro-

duced is provided near any switch which energizes the x-ray tube.

(4) A readily visible indicator is provided near the source housing to signify when the x-ray tube is energized or the radioactive source port is open.

(5) A radiation warning label is present near x-ray ON switches and on the control panel of a system containing a radioactive source.

(6) A label specifying the nuclide and its activity on a stated date is attached to the housing of a radioactive source.

(7) A description of the recommended operational and alignment procedures is present.

In addition, for enclosed beam systems only, the surveyor *shall* ascertain that:

(8) The *exposure* rate 5 cm from the protective chamber or chambers (enclosing the source, sample, and measuring system) cannot exceed 0.25 mR/h during normal operation.

(9) Access ports to the sample chamber are provided with interlocks of fail-safe design that prevent x-ray generation or entry of the x-ray beam into the chamber when any port is open.

In addition to items (1) through (7), for open beam systems only, the surveyor *shall* ascertain that:

(10) The *exposure* rate at 5 cm from the x-ray tube housing with all shutters closed cannot exceed 2.5 mR/h during operation at the maximum rated power and tube kilovoltage.

(11) Each port of the radiation source housing is provided with a shutter that can be opened only when accessory apparatus or a collimator is coupled to the port.

(12) A guard or interlock is provided which prevents exposure of any part of the body to the primary beam.

(13) Shutters at unused ports are secured to prevent casual opening and all shutters are provided with a fail-safe indicator of open/shut status.

Survey instruments used *should* be suitable for measurement of small, non-uniform x-ray fields with photon energies in the range of 5 to 100 keV. Corrections for energy dependence and/or size of the detector should be applied when indicated by the characteristics of the x-ray field being measured. Energy dependence is discussed in Section 5.1.1.1 and a method for detector size correction is described in Section 5.1.1.2. The area over which exposure rate is averaged should not exceed 1 cm^2. Ionization chambers with thin windows (1.0 mg/cm^2) are recommended for measurements in the useful beam and in stray radiation fields.

For the detection of small beams that penetrate defects in the shielding, photographic films, thin-walled geiger tubes, small scintillation detectors, and semiconductor detectors and fluorescent screens are useful.

Measurements *should* particularly be made of any leakage at closed ports, at the camera or diffractometer, at the beam stop, at any viewing device, and at each port when each shutter is open.

3.3.3 *Equipment Producing Unwanted X Rays*

Any device in which charged particles are or may be accelerated through potential differences exceeding a few thousand volts may be a source of x rays. Electronic equipment operating above 10 kVp *shall* be considered to be capable of producing radiation, and *shall* be surveyed as part of the design and production testing. High-voltage rectifiers, power amplifiers, television receivers, some microwave generators, electron microscopes, high-voltage vacuum switches, and vacuum condensers are examples of such equipment. Unless the equipment has been shown to require no radiation control measures under all possible conditions of operation, subsequent surveys during use *should* be made.

The normal operating condition at which the radiation output is maximum *should* be determined and the survey conducted under this condition. Measurements *shall* be made at adjacent positions that may be occupied during operation, particularly those closest to the apparatus. A counter-type survey instrument *should* be used to locate positions of significant *exposure* and the sources of radiation. Photographic film may be used to map the relative radiation intensities and to locate narrow beams that otherwise might be overlooked; however, quantitative analysis requires knowledge of the x-ray energies and the energy dependence of the film response. Quantitative measurements *should* be made with an ionization chamber suitable for low x-ray energies.

3.3.3.1 *Rectifiers, Power Amplifiers, and Thyratrons.* In high-voltage circuits, x rays may be emitted by rectifier tubes, power amplifier tubes, and thyratrons. Radiation levels up to 3000 mR/h have been measured near hydrogen thyratrons used in radar transmitters. Under normal conditions the voltage across a rectifier during the conducting phase is so low that the x rays produced are absorbed in the glass envelope of the tube. However, reduction in filament current may significantly increase the voltage drop across the tube and result in the production of more penetrating x rays. During the non-conducting phase when the voltage across a rectifier or thyratron is high, radiation may be produced by acceleration of charged particles produced in the gas in the tube, and of electrons emitted by hot anodes or extracted by cold emission from tube structures. Usually the radiation emission is highly localized. An ionization chamber rate meter is recommended as the measuring instrument but small beam corrections are usually necessary (See Section 5.1.1.2).

3.3.3.2 *Direct-View and Projection-Type Television Receivers.* The operating voltage of most picture tubes used in direct-view television receivers (home or industrial) is from 15 to 30 kVp. The average exposure rate over any 10 cm^2 area 5 cm from the surface of any home television receiver *shall not* exceed 0.5 mR/h when the receiver is operated under those conditions producing the maximum x-ray emission (NCRP, 1968b; CFR, 1976a). Home television receivers need not be surveyed if a recognized laboratory has certified that prototypes of the set conformed to the above recommendation. However, the possibility of radiation exposure to personnel during testing and servicing *should not* be overlooked.

Theater projection-type television tubes operate at voltages as high as 80 kVp and produce radiation levels significantly higher than direct-view types. Therefore, this type of receiver *shall* be surveyed to assure protection of operating and service personnel (DHEW, 1970a).

When a survey is required, measurements *should* be made in the vicinity of the high voltage rectifier tubes, the shunt regulator tubes, the viewing screen, and the funnel and base of the picture tube (DHEW, 1968; 1970b). For rapid location of the areas with the highest exposure rate, the use of a G.M. ratemeter instrument is recommended (Stoms and Kuerze, 1968). For measurement of exposure rates a ratemeter instrument of the ionization chamber type with a reasonably flat energy response to x rays generated at 10-40 keV (HVL 0.1-2 mm Al) (Els, 1971), a sensitivity in the order of 1 mR/h full scale, and an effective detector area of less than 10 cm^2 *should* be employed.

During the survey of a color televison receiver it *should* be noted whether or not there are warnings regarding possible x-ray emission if the high voltage is adjusted to an excessive value and if the cover to the high voltage enclosure is not closed.

3.3.3.3 *Microwave Equipment.* Microwave generators such as klystrons, magnetrons, traveling-wave tubes, and other radio-frequency (rf) tubes operating at peak voltages exceeding 15 kV may emit significant amounts of x radiation in addition to their production of rf fields (Lehman, 1970). The hazards associated with rf radiation are outside the scope of this report. Recommendations regarding x-ray surveys of microwave equipment have been made by the JEDEC Electron Tube Council (Electronic Industries Association, 1972).

X-ray leakage from electron beam microwave generators most frequently occurs from the gun, including the cathode bushing, the collector ends of the tube, the rf output window, and the anode walls. The radiation intensity is usually greater when the rf drive is applied to the device. The effective photon energy may range from 7 to 100 keV depending on the anode voltage employed, but commonly it is less than 25 keV.

The radiation survey of such equipment is similar to that of an industrial x-ray installation except for the following considerations:

(1) The presence of intense rf fields may cause erroneous readings in a survey instrument unless adequate rf shielding is provided (Bradley and Jones, 1970). The electrical shielding *should* be chosen so as not to increase significantly the energy dependence of the instrument response. Transparent conductive cloth (e.g., silver-coated nylon marquisette) is suitable for this purpose (Properzio, 1970).

(2) Microwave generators frequently emit radiation in pulses of microsecond duration. Since extremely high instantaneous exposure rates are possible, care *shall* be taken to assure that the survey instrument responds properly to the high exposure rate involved (e.g., that voltage saturation is maintained in ionization chambers, especially those of large diameter). High rate response is discussed in Section 5.1.1. Counter-type detection instruments *should not* be used to measure pulsed radiation fields.

The recommended instruments for survey use are ionization chambers of the ratemeter or integrating type with minimal or known energy dependence particularly in the range 7 to 25 keV (HVL 0.1 mm to 1.0 mm of aluminum). Measurements *should* be made at distances from accessible surfaces of the microwave device which are several times the dimensions of the chamber; 30 cm is a reasonable distance. X-ray film of the medical no-screen type or industrial type is also recommended to determine the spatial pattern of the radiation and particularly to locate leaks of very small area. Metal filters over the film may be used for photon energy estimation. The film may need to be shielded from thermal radiation.

Radiation surveys *should* be made with the microwave tube operating at full rf power. It is possible that x rays may be produced in the absence of an rf output or when the electron beam is defocused; therefore radiation surveys *should* also be made under these conditions.

The surveyor *shall* determine whether or not radiation warning signs and caution notices are required under applicable regulations and whether increased radiation shielding is necessary.

3.3.3.4 *Electron Microscopes.* These units usually operate at 50 to 100 kVp and 5 to 500 μA. Recently electron microscopes operating up to 1000 kVp have been produced. Radiation levels at the electron gun, camera, ports, and other accessible locations *should* be determined. Radiation levels at the high voltage power supply unit *should* also be determined. Ionization chamber instruments suitable for measurements of low exposure rates in low-energy x-ray fields *should* be used (Parsons *et al.*, 1974).

3.3.4 *Accelerators and Neutron Generators*

The following discussion relates to devices that accelerate charged particles such as electrons, protons, deuterons, and helium ions to energies greater than about 1 MeV, and to neutron generators that operate with accelerating voltages down to about 150 kV. These devices include cyclotrons, betatrons, linear accelerators, Van de Graaff accelerators, and Cockcroft-Walton neutron generators. This discussion does not apply to medical x-ray generators operating up to 10 MeV since the survey of these installations is covered in Section 3.3.1.4.

Electron and heavy particle accelerators produce intense beams and radiation fields that contain x-ray, gamma-ray, and particulate components having a wide spectrum of mass, charge, energy, and intensity. Irradiation of individuals working in the vicinity of an accelerator during operation may arise from direct or scattered beams, from radiation beams modified by shielding, from tritium liberated from targets, or from radioactivity induced in irradiated materials that may persist when the accelerator is not in operation. Some of the radiation fields encountered may have high intensities, yet may not be detected easily; for example, narrow pencils of radiation escaping through cracks in the shielding. Special measurement problems occur around accelerators that produce high-intensity pulsed radiation fields. The choice of survey instrumentation and the application of biological and operational dose-modifying factors require special consideration. Because of the complexity of these problems, supervision by qualified experts (see Introduction) is required during planning and construction as well as during the operation of such installations. Accelerator installations operating at energies in excess of 100 MeV are considered beyond the scope of this report.

Valuable information for persons involved in the radiation protection problems oι accelerators and neutron generators is contained in NCRP

Report No. 38 (NCRP, 1971a), NCRP Report No. 23 (NCRP, 1960), NCRP Report No. 25 (NCRP, 1961a), NCRP Report No. 51 (NCRP, 1977), in the NCRP report on small neutron generators (in preparation), and in NBS Handbook No. 107 (ANSI, 1969c). Further information is available from Brobeck *et al.* (1968) and from Patterson and Thomas (1973).

Special monitoring techniques useful around these machines are discussed in this Report and in NCRP Report 51 (NCRP, 1977). Some of the instruments and methods described in these handbooks are highly specialized and require a high degree of technical proficiency for their calibration and use. This section will emphasize methods that permit the performance of surveys with less sophisticated instruments.

3.3.4.1 *Inspection.* In addition to the general inspection requirements outlined in Section 3.2, the survey *shall* determine that:

(1) "Emergency off" switches are installed that will suspend operation so as to prevent radiation production and that will require local resetting before radiation production can proceed.

(2) Such switches are present in sufficient number in the irradiation area and in all other high radiation areas, that at least one is visible from any part of these areas and readily accessible over free and unencumbered passages.

(3) Clearly visible warning lights are displayed on the control panel, at the entrance to high radiation areas, and *in* such areas, while radiation is being produced. The nature of all warning lights (color, flashing, rotating) and their consistency with American National Standard Z53.1-1971 should be noted (ANSI, 1971).

(4) A warning by light of another color or by clearly audible sound indicates the activation of the time-delay circuit prior to turning on the radiation, and continues for the entire duration of the time-delay.

(5) The duration of the time-delay is sufficiently long to permit personnel to escape or activate emergency-off switches.

(6) Doors from enclosed high radiation areas can be opened from within the enclosure, even in the event of power failure.

(7) Prominent warning notices are present that concisely indicate the meaning of automatic warning devices.

(8) Area monitors are provided in radiation areas that can be occupied by personnel, are functioning properly, and warn of radiation levels above a predetermined limit.

(9) The ventilation system for the irradiation room is interlocked with the control circuits of the equipment (in those installa-

tions where special ventilation is required during equipment operation).

(10) All entrances into the irradiation room or other high radiation areas are provided with barriers equipped with interlocks that are not dependent on the operation of a single circuit, and that will interrupt radiation production when the barrier is opened.

When accelerators or neutron generators are located in medical therapy installations, the surveyor *shall* also ascertain that:

(11) A device is provided that terminates exposure after a preset dose or a preset time.

(12) A back-up timer or dosimeter is provided that will turn off the radiation should the preset exposure control fail to operate.

(13) The indicator of radiation output is interlocked with the control circuits of the accelerator so that production of radiation will be prevented if the indicator is not operating.

(14) In accelerators supplying both electrons and x rays, an interlock is provided and operating to prevent inadvertent electron exposure when x-ray exposure is indicated.

(15) Provision is made for observation and oral communication with the patient from the control station during treatment.

Operating procedures vary with each accelerator installation, but some basic practices for reducing the possibility of accidental human exposure apply to all of them. A surveyor *shall* investigate the operating procedures to determine that:

(16) The irradiation room and all other high radiation areas are inspected by the operator before turning the radiation "ON".

(17) Apparent malfunctioning of any irradiation control and monitoring equipment is promptly investigated and corrected.

(18) An up-to-date log book is kept that records calibrations and changes in the equipment that affect operating characteristics.

(19) Access to the irradiation room, after the accelerator has been turned off, is delayed for the period of time that may be required for appropriate reduction of induced radioactivity or ozone concentrations.

3.3.4.2 *Measurement of the Useful Beam.* The intensities of all the radiations produced by an accelerator depend strongly on the type of particle accelerated, the target material, the beam energy, and the output or beam current. Therefore, all of these characteristics prevailing at the time of the survey measurements *shall* be noted.

The useful beam may be an electron beam, *bremsstrahlung* produced in an internal target, a beam of heavy charged particles, or a neutron flux. Accelerators are usually equipped with monitors that permit the determination of output and energy of the useful beam with sufficient accuracy for radiation hazard evaluation. With neutron generators, the current of accelerated charged particles is usually monitored. However, since the condition of the target affects the neutron flux density, a new survey *should* be made whenever the target is replaced.

Ionization chambers are most commonly employed for monitoring the useful beam in accelerators of heavy charged particles and in electron accelerators used for medical therapy or industrial radiography. Secondary emission monitors are frequently used with high-current pulsed electron beams. For electron or x-ray beams the monitor is usually calibrated against a secondary standard for absorbed dose in tissue. This may be a thimble ionization chamber inserted in a block of polystyrene at a depth corresponding to the maximum of the depth dose curve. Precautions that may be taken to assure reliable measurements with thimble ionization chambers, when these are used with high-energy pulsed radiations, are discussed in Section 5.1.1.1.

The dimensions and locations of an external beam of charged particles can be obtained by exposure of a radiographic fluorescent screen viewed remotely, by exposure of x-ray film with or without subsequent processing, polyvinyl chloride sheets, thin quartz, glass or aluminum plates, or by autoradiographic studies of radioactivity induced in suitable foils. If the beam is defocused or displaced, it may produce intense beams of radiation at unexpected locations.

3.3.4.3 *Measurements of Stray Radiation.* Stray radiation may include leakage of the primary radiation beam, scattered radiation, and x rays, gamma rays, and neutrons generated by the primary beam. Except at the highest energies used in physical research, an adequate concrete shield for x or gamma rays usually reduces the fast neutron background to well below permissible levels. However, if access to the irradiation room is through a shielded lead or steel door, a high background of fast neutrons will often be encountered in its vicinity.

Survey of the radiation due to electrons, *bremsstrahlung*, or gamma rays *should* be carried out with ionization chambers. Counter-type instruments have a limited usefulness for measurement of these radiations from pulsed accelerators except for the measurement of radioactivity. A wide-range survey instrument (e.g., with several decades of a logarithmic scale) is very useful.

The most useful neutron detector for the survey of occupied areas around accelerators is considered to be a BF_3 proportional counter with a removable moderator made from hydrocarbon material, because it has a high sensitivity to both thermal and fast neutrons. A paraffin moderator 2½ inches thick leads to an approximately constant sensitivity (± 20 percent) to neutron flux density for fast neutrons in the energy range from 0.1 to several MeV. Such a counter *should* be calibrated against a standard neutron source. Commercially available nuclear track film is recommended for the evaluation of the primary barrier as a neutron shield. Whenever feasible, more than one method should be used to evaluate neutron doses, so as to give assurance as to the validity of the survey. A tissue-equivalent ionization chamber permits an evaluation of the dose in mixed radiation fields of gamma rays and neutrons, and has the advantage that it can be calibrated with gamma rays. However, an evaluation of the dose-equivalent requires additional information on the relative neutron and gamma-ray components. An upper limit to the dose equivalent is obtained if one can reasonably assign a quality factor of 10 to the mixed radiation present. A tissue-equivalent chamber is thus particularly useful in the survey of heavy particle accelerators where exposure to fast neutrons predominates. A detailed discussion of neutron instrumentation appears in Section 5.1.2.4.

3.3.4.4 *Induced Radioactivity.* The residual induced radioactivity in the irradiation room after the accelerator has been turned off may be a significant problem with high power industrial electron accelerators operating above 10 MeV (NCRP, 1977), and with all heavy particle accelerators. Typically, in electron accelerator installations, positron-emitters are induced; these can be detected either directly or through the associated 0.51 MeV annihilation gamma rays by standard beta-gamma survey instruments. Radioactivity will be induced in air (nitrogen-13 and oxygen-15 with half-lives of 10 and 2 minutes, respectively) in electron accelerator installations operating at high power levels and at energies higher than the photodisintegration thresholds for these reactions (10.5 MeV and 15.6 MeV, respectively) (Brobeck *et al.*, 1968; Barbier, 1969). In heavy-particle accelerator installations, high radiation intensities due to induced radioactivity are usually present in the vicinity of the target, the injector, the pumps, and other parts where the presence of personnel may be required for maintenance or adjustment purposes. Radiations from radioactive isotopes of copper, zinc, and nickel are frequently present. *Exposure* rates from gamma rays near the vacuum tank of a cyclotron may range up to 10 R/h immediately after shutdown (Cook, 1963). In addition, radioactive contamination may be present in airborne particulates or as dust in the irradiation area.

Measurements of ambient radiation levels *should* therefore be made before or during each entrance to the irradiation area after accelerator shutdown. During an initial survey, measurements of airborne activity *should* be made both during and after the operation of the accelerator. The surveyor *should* consider the adequacy of the ventilation and of any delay period imposed before entrance to the irradiation room. He *should* also consider whether continuous air monitoring is necessary. He *should* also review the precautions taken regarding the removal, handling, transfer, and storage of the target and other possibly contaminated components. Periodically, wipes of the floor and external surfaces near the target of the accelerator *should* be made to detect the build-up of long-lived removable activity.

In the survey of residual radiation intensities around a cyclotron the surveyor *should* note whether the magnet is energized, since its magnetic field restricts the beta radiation to the immediate vicinity of the cyclotron. Normally, rate meters cannot be used in the strong magnetic field close to the cyclotron magnet; however, film, pocket ionization chambers, or thermoluminescent dosimeters are useful in relating exposures in locations inside the magnetic field to those outside the field.

3.3.4.5 *Tritium.* Particle accelerators, particularly those of the Cockcroft-Walton type, may be used for the production of neutrons through the (d,t) reaction with a tritiated metal target. Approximately one tritium atom is released for each bombarding deuteron and most of these atoms are either trapped in the pumping system or released through the vacuum system exhaust (Nellis *et al.*, 1967; Watson *et al.*, 1969). Pump oil *should* be assayed for tritium content periodically, particularly before the pumps are serviced. Tritium levels *should* also be monitored when the vacuum system is opened and wipe tests should be made on parts to be handled. The exhaust from the pumping system *should* be monitored and the tritium concentration compared with permissible concentration limits at the point of release. Tritium concentrations in storage areas for used and unused tritiated targets *should* also be measured and consideration given to the adequacy of tritium containment and ventilation of the storage area. Methods for the measurement of tritium contamination in gaseous form are discussed in Section 5.2.2.1 and in the form of liquid samples or wipes in Section 5.2.3.4 and 5.2.3.5.

3.3.4.6 *Ozone.* High intensity electron beams passing through the air may produce ozone in toxic concentrations. The threshold limit value is 0.1 parts per million (ACGIH, 1971). A method for estimating the ozone concentration is given in NCRP Report No. 51, Appendix I (NCRP, 1977). This hazard is reduced by providing adequate ventilation. Calculations have been made of the time required before the concentration is reduced to the threshold value as a function of beam current, path length, irradiation time, and exhaust rate (Brobeck *et al.*, 1968; Brynjolfsson and Martin, 1971). Where such calculations indicate a possible hazard, and particularly if the odor of ozone is noticeable, the concentration *should* be measured with a suitable instrument. The surveyor *should* consider whether improvements in the air flow or location of air exhaust vents are needed.

3.3.4.7 *Auxiliary Equipment or Accelerators Partially Energized.* During maintenance and tune-up, it is often necessary for personnel to work in areas from which they are normally excluded. Before and during this work a survey of occupied areas *shall* be made. The surveyor *should* ascertain whether personnel are wearing appropriate monitors, such as pocket chambers, that permit a rapid evaluation of the exposure. Prevailing radiation conditions *shall* be measured. Although the electron or ion source is usually turned off during such work, some of the circuits are operative and may cause radiation to emanate from unsuspected locations. Moreover, auxiliary high voltage vacuum tubes, such as klystrons and rectifiers, may be sources of x rays (see Section 3.3.3).

Electrons released by various mechanisms within the machine can produce x rays at the positive high voltage terminals whenever the accelerating mechanism is "ON". A possibility that *should* be considered with positive ion accelerators, particularly with Van de Graaff machines, is that the accelerating voltage may be accidentally reversed. Electrons generated within the tube will be accelerated toward the target end of the machine and the emerging electrons or x rays may produce a serious hazard.

Malfunctioning magnets or poor vacuum in part of the system can cause the beam to strike the walls of the vacuum chamber and other structures and generate x rays, while the primary beam monitor indicates reduced intensity or absence of the primary beam. When deuterons are accelerated they may become imbedded in materials they strike and act as targets for other deuterons in the high yield d(d,n) reaction. Control and defining slits as well as other parts of the vacuum chamber of the analyzing magnet may become such secondary sources, unless they are heated to drive off the deuterium stopping in them.

3.3.4.8 *Patient Treatment Location.* In accelerator installations used for radiotherapy, measurements *shall* be made at the treatment site to detect flaws in the shielding of the machine which might cause the patient to be exposed to narrow beams of intense leakage radiation and to evaluate the dose contributed to the patient by stray radiation. ICRP Publication 15 (ICRP 1970) recommends that the dose rate from leakage radiation transmitted through the beam collimation device *shall not* exceed 2 percent of the dose-rate in the useful beam at the same distance from the source. Leakage radiation reaching the patient at any other point *shall not* exceed 0.1 percent of the useful beam. In the medical application of cyclotrons or synchrotrons the beam normally passes through an opening in the primary barrier; the adequacy of this barrier *shall* be verified.

The stray radiation in electron or x-ray therapy installations consists of scattered electrons in the immediate vicinity of the primary beam, high energy *bremsstrahlung*, and fast neutrons, if the primary beam energy exceeds 10 MeV (Berger and Seltzer, 1970). The neutron yield increases rapidly with electron energy up to about 20 MeV. The machine outputs used in therapy are such that induced radioactivity is not a problem at the treatment site. For survey measurements of electrons and *bremsstrahlung*, ionization chambers of the rate or integrating type *shall* be used. In the former situation, the meter *shall* be read by remote methods.

The neutron dose equivalent can be calculated from the neutron fluence and the neutron spectrum most conveniently by use of the data appearing in Appendix B of NCRP Report No. 38 (NCRP, 1971a); alternatively, the dose equivalent rate can be calculated from Table 2 in the same report. The measurement of the fast neutron fluence or flux density in the immediate vicinity of a pulsed photon or electron beam is a difficult problem. This is restricted in practice to neutron foil detectors (used as threshold detectors or in a paraffin moderator), fission track detectors, and silicon diode dosimeters since they offer high discrimination against gamma rays[12] and their sensitivity is adequate for this work (Axton and Bardell, 1972; Wilenzick *et al.*, 1973). The utilization of foils and fission track devices for the detection of neutrons is discussed in Section 5.1.2.4. Spectra of fast neutrons produced by photodisintegration from 20 to 35 MeV *bremsstrahlung* have been measured with nuclear emulsions; the average energy of photoneutrons produced in lead has been estimated at 1.7 MeV. The published spectra generally have a peak at approximately 1 MeV and a high energy tail and can be used to approximate the neutron spectrum in the vicinity of the patient.

A correction may be necessary for photoneutrons generated within a thick paraffin moderator by *bremsstrahlung* of energy above approximately 20 MeV due to the $^{12}C(\gamma,n)$ reaction. The neutron flux density generated within a patient may be of the same order of magnitude as the incident neutron flux density. A large paraffin

[12] None of these devices is completely insensitive to photons with energies above their threshold for nuclear reactions (typically 8 MeV). For example, high-energy photon reaction cross-sections are approximately 100 times smaller than fast neutron cross-sections in silicon diode dosimeters and therefore in a field where the photon to neutron fluence is high, a significant part of the response will be due to photons (McCall, 1975).

phantom in which neutron activation foils are imbedded can be used to estimate its magnitude.

3.3.5 *Sealed Radioactive Sources*

The activities in sealed sources range from megacuries for food processing and experimental purposes; kilocuries for teletherapy; curies for industrial radiography, neutron production, and static elimination; millicuries for brachytherapy; to submicrocurie levels for use in gauges, luminous devices, and gas-filled electronic tubes.

There are two important characteristics of sealed sources that require special survey procedures. These are:

(a) the possibility of serious airborne or surface contamination should the integrity of the source container fail;

(b) the inability to turn the radiation source off.

Leakage may result from mechanical damage, from radiation damage, or from chemical corrosion of the sealed source capsule. It also may be due to mechanical stresses in the capsule caused by large temperature variations, or by high gas pressures generated inside the capsule by gaseous daughter products of the parent nuclide, or by gaseous chemical reaction products.

The radiation from sealed radioactive sources may consist of gamma rays, x rays, beta rays, alpha rays, or neutrons. Some sources emit several types of radiation. Beta-ray sources may produce x rays (*bremsstrahlung*); neutron sources of the (γ,n) and the (α,n) type may release gamma rays as well as neutrons. In the case of photoneutron (γ,n) sources employing radium, antimony-124, etc., the dose rate from the gamma rays may exceed that from neutrons by as much as four orders of magnitude. However, if these sources are heavily shielded with a high atomic number material such as lead, the gamma rays may be practically eliminated but the neutron dose rate may still be appreciable. The (α,n) sources also produce gamma rays as well as neutrons, but usually the flux densities are comparable. This means that the dose equivalent rates from the neutrons are usually greater than those from the gamma rays.

Protection against radiation from sealed gamma-ray sources is extensively discussed in NCRP Reports No. 40 (NCRP, 1972) and No. 49 (NCRP, 1976b), from neutron sources in NCRP Report No. 38 (NCRP, 1971a), and from industrial beta-ray sources in NBS Handbook 66 (1958). In this Report, information on radiation surveys contained in the foregoing reports has been collected and expanded.

3.3.5.1 *Inspection.* In addition to the general inspection requirements outlined in Section 3.2, the surveyor *shall* determine that:

(1) The source container is labeled with the accepted radiation warning symbol, the amount and type of radionuclide, and the date of measurement.
(2) The restrictions (e.g., locks) incorporated in the device, storage container, or storage area assure that the source is not accessible to unauthorized persons.
(3) The procedures and records designed to prevent loss of the source(s) are adequate.
(4) For sealed sources in protective source housings, the shutter or on-off mechanism and corresponding indicators are reliable and adequate.
(5) For sources in protective housings where there is no shutter or on-off mechanism (such as some gauges and static eliminators), there is effective means to prevent access to the useful beam.

For apparatus in which sealed sources are removed from their protective housing during use, the surveyor *shall* determine that:

(6) There is a warning tag or label on the source or source holder.
(7) When the source is used in an unshielded area, the perimeter of the restricted area is clearly marked and under surveillance.
(8) Adequate procedures have been prescribed for withdrawal, use, and return of the source.

3.3.5.2 *Measurement of Radiation Fields.* Large sealed sources such as teletherapy sources, radiography sources, and neutron sources frequently emit beams of radiation and, therefore, surveys around them *should* be conducted in the manner recommended in Section 3.2 and in particular in Sections 3.3.1 and 3.3.2 for x-ray installations. Measurement of the useful beam, scatter, and leakage radiation levels *shall* be included.

Where sealed sources are used in open, unshielded areas, measurements *shall* be made to assure that restricted areas around the sources are sufficiently large to limit the dose equivalent received by persons outside the perimeter to MPD values.

Where sealed sources, prior to use, are moved out of a shielded storage container, e.g., to the end of a remote control cable, measurement or inspection after each use of the device *shall* be made to provide assurance that the source has returned to its stored position in the container.

Around electron tubes, gauges, and luminous devices containing radioactivity in microcurie amounts, a potential for significant external exposure may exist only when a large number of them are stored together or used in one piece of equipment. Under these circumstances measurements *should* be made.

Immediately after the installation or replacement of a large sealed source, such as a teletherapy or radiography source, the leakage radiation transmitted by the housing with the source in the "OFF" position *shall* be measured at a number of points, either on the exterior surface or at a defined distance from the surface (or source). A sufficient number of measurements *shall* be made to establish the average and maximum radiation levels. These measurements *shall* be made and evaluated with reference to the limits stated in the relevant recommendations contained in NCRP Report No. 33 (NCRP, 1968a) and NBS Handbook 114 (ANSI, 1975).

In teletherapy units, the leakage radiation transmitted by the housing and by the collimating device with the source in the "ON" position *shall* also be measured. Such measurements are normally made on prototype units and when this has been done they need not be repeated.

3.3.5.3 *Leak Tests.* The purpose of leak testing of sealed sources is to establish whether or not the capsule seal has failed, and not to characterize the extent of the hazard produced by the failure. Any amount of contamination measurable by commonly available counting apparatus is symptomatic of a failure.

All sealed sources *shall* be tested for leakage:

(1) before initial use;
(2) at intervals not exceeding six months, when the half-life exceeds 30 days;
(3) whenever damage or deterioration of the capsule or seal is suspected;
(4) when contamination of handling or storage equipment is detected.

Because of the rapidity with which gases escape from a leaking gaseous sealed source and diffuse into the air, periodic leak tests are of little value. Where large quantities of radioactive gas are stored or used, and air contamination due to leakage may produce significant fractions of the Maximum Permissible Dose to individuals in the vicinity, the survey *shall* ascertain that continuous monitoring is present.

Leak tests of solid or liquid sources may be made by a number of different methods, all of which depend on the detection of contamination on or near the source. These are reviewed in NCRP Report No. 40, Section 4.5 (NCRP, 1972). The method adopted depends on the size and accessibility of the source and the nature of the radionuclide.

A small sealed capsule may be tested by washing for a few minutes in a detergent solution. An aliquot of this solution *should* then be counted with a suitable detector (see below). An absorbent liner in the storage container normally in contact with the source will also reveal leakage if it is contaminated.

Leakage of radon-222, the gaseous daughter product of radium-226, may occur from a radium capsule through small cracks not permitting leakage of the radium element. Radium sources *should* therefore be leak-tested by detecting radon leakage either directly or through the alpha or beta-gamma activity of its daughters. Several methods are listed in NCRP Report No. 40 (NCRP, 1972) and described by Wood (1968). The presence of abnormal radon concentrations in a radium safe may be detected by periodic or continuous air sampling (Section 3.2.5).

Alpha and low energy beta sources require especially thin-walled capsules that must be handled carefully to avoid rupturing the seal. Leak tests *should* be made by carefully wiping the surfaces with absorbent cotton tipped swabs (Caruthers and Maxwell, 1971; Robertson and Randle, 1974).

Detection of contaminants on the housing or surface of a neutron source may not indicate source leakage, but may be due to induced activity. Confirmation of leakage may require identification of the contaminant.

Leak tests of devices from which the encapsulated source cannot be removed *should* be made by wiping the accessible surfaces of the source capsule or the accessible surface of the device nearest to the storage position of the source. The method of wiping *should* be chosen so as to avoid excessive personal exposure during this procedure. In a teletherapy unit, the accessible surface may be the inside surface of the collimator or the accessible part of the closed shutter. The wipe may be made with a small piece of filter paper. Wipes and suitable controls *should* then be counted.

The apparatus used for counting in a leak test *shall* be capable of detecting 0.005 μCi (11,000 dis/min) of the specific nuclide. In the case of radon and its daughter products the minimum detectable activity *shall* be at least 0.001 μCi. Suitable apparatus for counting gamma-ray activity is a scintillation well counter and for beta-ray activity a thin-window Geiger-Mueller counter, gas flow proportional counter, or a liquid scintillation counter. Tritium contamination should be counted by use of a gas-flow counter or liquid scintillation counter. Alpha activity may be detected with suitable sensitivity by thin-window or gas-flow proportional counters, semiconductor detectors, or alpha scintillation counters.

Leak tests that reveal the presence of 0.005 μCi or more of removable contamination, or, in the case of radium, the escape of radon at the rate of 0.001 μCi or more per 24 hours, *shall* be considered evidence that the sealed source is leaking. It *shall* be withdrawn from use and sealed in a separate container. The source *should* be returned to the supplier or sent to some other qualified person for repair or disposal. The source container or housing may require decontamination. A survey *should* be made to determine the extent of contamination of the installation. Regulations may require the notification of the regulatory agency when the contamination measured in a leak test exceeds a stated quantity.

3.3.6 *Unsealed Radioactive Sources*

The distinguishing characteristic of installations using unsealed sources is their potential for producing contamination within the installation and in the public domain. Unsealed sources are the principal cause of internal exposure in humans. Alpha particles, beta particles, gamma rays, and possibly neutrons (e.g., from α-n reactions) may be emitted over wide ranges of energies. Emitters of gamma rays, high-energy beta rays, and neutrons are of concern both as sources of external and internal exposure. Emitters of alpha particles and low-energy beta particles are an unimportant source of external exposure, but when deposited in the body they become important sources of internal exposure.

The dose equivalent produced by a radioisotope deposited in the body depends upon the type of radiation emitted, the physical half-life, the physical and chemical form and its solubility, and the metabolic properties of the material. One index of relative hazard is provided by the maximum permissible concentrations and maximum permissible body burdens listed in NCRP Report No. 22 (NCRP, 1959).

Generally, area surveys in installations employing radionuclides *should* include the detection and measurement of surface, airborne, and waterborne activity as well as the measurement of the radiation fields produced. Methods are presented for measurement of radiation fields in Section 3.2.4 and of surface and airborne contamination in Section 3.2.5.

3.3.6.1 *Investigation.* The investigational phase of area surveying is of particular importance for unsealed sources. All subsequent steps depend on the information obtained regarding the particular radionuclides, quantities on the premises, and the specific uses made of them. In addition to the general investigational requirements outlined in Section 3.2., the following information *should* be obtained:

(1) The inventory of and location of radioactive materials and, to the extent practicable, their physical and chemical form.
(2) The uses of the radioactive materials.
(3) The identity of the person(s) responsible for supervision and control of the radionuclide inventory.
(4) A copy of the local rules and procedures used for control, and for instruction of persons handling unsealed sources.

3.3.6.2 *Inspection.* The physical safeguards and procedural controls employed in installations having unsealed sources will vary widely with the quantity, type, and form of the radionuclides and with the function of the installation. Therefore, not all of the items specified below will be relevant. During the inspection of physical safeguards the surveyor *shall* determine, when relevant, that:

(1) Hoods are provided and are adequate, paying particular attention to ventilation, shielding, and confinement. A linear air flow rate of 100 to 200 feet per minute at the face of an open laboratory-type hood is recommended, depending on the activity and radiotoxicity of the material.
(2) Glove boxes are provided, paying particular attention to the shielding, the protection afforded by the gloves, and the airlocks for transfer of materials.
(3) Long-handled tools and remote-handling devices are available and are used.
(4) Protective clothing and respiratory protective devices are provided.
(5) Facilities for handling, storing, and disposing of liquid and solid wastes are provided.
(6) Storage containers and areas are shielded and posted.

During the inspection of procedural controls the surveyor *shall* determine that:

(7) Source containers are labeled with the accepted radiation warning symbol, the amount and identity of radionuclide, the physical and chemical form, and the date of measurement.
(8) Adequate procedures are used for isolating, decontaminating, or disposing of contaminated material, equipment, and protective clothing.
(9) There is a system of procedures and records to prevent unauthorized use or loss of the materials.

3.3.6.3 *Waste Disposal Systems.* Under prescribed conditions, low-activity liquid waste may be released into the sewage system, and solid wastes may be incinerated or buried. Recommendations for disposal of radioactive materials are contained in NCRP Reports Nos. 9, 12, 16, and 30 (NCRP 1951b, 1953, 1954b, and 1964). Applicable regulatory codes *should* be consulted for limitations; e.g., Code of Federal Regulations, Title 10, Sections 20.301 - 20.305 (CFR, 1976d).

Sinks, drains, traps, pipelines, and storage tanks used in the disposal of liquid radioactive wastes *should* be surveyed periodically. Sites at which radioactive materials may reconcentrate *should* be surveyed. When drains used for disposal are opened for repairs, survey measurements *should* be made promptly. Such surveys normally consist of measurements of radiation and contamination levels, but in some situations radioactive assays of liquids or sludges may be necessary. If so, an effort *should* be made to secure a representative sample either by stirring static liquids or removing samples of liquid or sludge from several locations. Concentrations in a flowing liquid *should* be determined by sampling at regular time intervals. Evaporation of liquid samples to increase the concentration may be necessary where activity levels or instrument sensitivity are low. NCRP Reports No. 28 (NCRP, 1961b) and No. 30 (NCRP, 1964) and Sections 5.2.2 and 5.2.3 of this report discuss instrumentation for making these measurements and sampling techniques.

When waste is incinerated it is recommended that the stack effluent *should* be monitored, employing techniques such as those described in Section 3.2.5.2. Air monitoring provides the best estimate of the average air concentration at the point of release. In addition, periodic area surveys of rooftops and/or other local areas close to the base of the stack are recommended in order to detect build-up of activity through fallout or precipitation. At the present time, with the exception of light water power reactors, air concentrations of a mixture of radionuclides, averaged over a period of one year at the boundary of an installation, are limited by federal and local regulations to the non-occupational MPC_a for that mixture[13]. Alternatively, limitation of the air concentration can be achieved through control of the quantity of radionuclides incinerated.

A survey of a facility for incineration of radioactive waste *shall* include an evaluation of the concentration of radioactivity in the effluent stream to determine whether it conforms to applicable regulations. In a facility with effluent monitoring, records of effluent activity *should* be reviewed. In many facilities, the concentration of radioactivity in effluents averaged over one year is limited at the point of discharge to the non-occupational MPC_a. In such facilities, if only one radionuclide is released, the quotient of the total annual activity released and the total annual effluent volume can be compared with the MPC_a for that nuclide. If several radionuclides are released, the

[13] Before this limit is applied for releases to an outdoor area, consideration *should* be given to the potential for reconcentration in foodstuffs through ecologic factors.

total activity of each radionuclide should be derived and the fraction of the MPC_a for each radionuclide computed (Section 3.2.5.2). The sum of the fractions *should not* exceed unity. If the mixture of radionuclides is unknown, comparison of the effluent concentration *should* be made with the non-occupational $MPCU_a$.

If effluent control is achieved through limitation of the activities incinerated, records of these activities *should* be reviewed. The average concentration of activity contained in the effluent *should* be estimated for each radionuclide, employing reasonable values for the volatile fractions (Bush and Hundel, 1973; Wollan *et al.*, 1971; Geyer *et al.*, 1956; AEC, 1970). The average effluent concentration *should* be compared with the MPC_a weighted as indicated above.

The disposal of incinerator ash to uncontrolled locations requires special permission from the regulatory agency. At present an acceptable criterion is the limitation of the average concentration of the radionuclides in the ash to less than the non-occupational MPC_w for the mixture which is present. The basis for this practice is that the concentration of radionuclides in run-off water, which could eventually enter the public water supply, will be much lower than the MPC_w. For these purposes the averaging *should* be limited to each batch of ash disposed. If the concentration exceeds the MPC_w values, the ash *should* be stored or removed in a sealed container to an approved disposal site. During the inspection phase of a survey, records relating to the activity in ash and methods of ash disposal *should* be examined. Methods of sampling, radioactivity measurement, and collection of ash *should* be reviewed.

When disposal is accomplished through burial, the specific procedures regarding depth, spacing, etc., *should* be examined to determine whether they conform to applicable regulations (e.g., Code of Federal Regulations, Title 10, Section 20.304) (CFR, 1976d). Burial containers and sites *should* be examined to determine whether they are properly marked and isolated. Radiation levels at burial sites for gamma emitters *shall* be measured.

3.3.7 *Reactors and Critical Facilities*

This section is intended to apply only to subcritical assemblies, and to critical assemblies or reactors with a power level limited to about one megawatt. Such devices are commonly used as sources of neutrons and ionizing radiations for research, training, and testing purposes. Reactors operating at power levels above about one megawatt are outside the intended scope of this report; the radiation monitoring and surveying programs for such installations *should* be directed by persons specially trained in reactor monitoring.

The protection problems encountered around reactors and critical facilities embrace most of the problems met in handling sealed and unsealed sources, and in operating accelerators. Therefore, much of the previous discussion on these subjects (Sections 3.3.4, 3.3.5, 3.3.6) applies here. However, there are several unique protection problems associated with the handling of fissionable materials and with the disposal of reactor wastes that require special surveying and monitoring methods and competent supervision.

First, unusual potential for exposure arises from the radiation fields which are mixed and often intense. Examples are leakage radiation through openings in the concrete shield or during the replacement of reactor fuel assemblies.

Second, the air may be contaminated by fission products and neutron-activated noble gases, and intense radioactivity due to massive neutron activation of the coolant or moderation systems and of corrosion products may be present. The added risk of internal exposure from such contamination must be considered in addition to the potential for external exposure from the radiation field.

Third, there is a potential for accidental loss of control over the fission rate with abrupt increase in the power level, that may result in greatly increased needs for radiation protection. Loss of shielding may result in very high external dose rates to persons in the vicinity, and loss of containment may result in the release of radioactive materials in quantities sufficient to pose serious internal and external exposure risks to persons in the buildings and surroundings.

Radiation protection in critical facilities is achieved principally by numerous types of physical safeguards including interlocks, alarms, automatic monitoring, barriers, and containment systems. Procedural controls are used to assure the maximum effectiveness of the physical devices and to maintain safe operating practices.

3.3.7.1 *Inspection.* In addition to the general inspection requirements outlined in Section 3.2.2, a radiation survey of a reactor or critical facility *shall* include a determination of the existence and correct functioning of:

(1) Instrumentation and personnel dosimetry devices for measuring a wide range of neutron, beta, and gamma-ray exposure.
(2) Detection systems capable of detecting an unplanned reactor excursion or a criticality accident.
(3) Monitoring systems for gaseous and liquid effluents.
(4) Fail-safe automatic alarm systems and signals for notification of all persons in the facility in general emergency situations.

(5) Interlocks, emergency shutdown, and alarms in special areas of high exposure risk.

In addition, the survey *shall* determine the existence of:

(6) Secondary shields for the coolant or radiation beams for experimental purposes, outside the primary core shield.

(7) Procedures for exclusion of personnel from high radiation areas during all phases of operation of the facility.

(8) Current records indicating areas where survey measurements have revealed significant leakage from the biological shields.

(9) Detailed emergency instructions and plans covering accidental criticality, environmental releases, power excursions, sabotage, and natural disasters.

3.3.7.2 *Surveys of Core Shielding.* The radiation emerging from the core of a reactor or critical facility during its operation consists of a mixture of gamma rays and neutrons. Gamma rays with energies from a few keV to many MeV and neutrons from thermal energies to many MeV are present. The typical fission neutron energy spectrum exhibits a maximum at about 0.75 MeV with an upper-level cutoff above 10 MeV; however, the typical reactor neutron spectrum is degraded in energy with an increase in the proportion of low energy neutrons. The radiation levels from the operating core are almost directly proportional to the power level, i.e., to the fission rate. The component of reactor radiation that is directly associated with the fission process can be turned on and off.

In some reactors, the core is immersed in a pool of water that serves both as the primary shield and the coolant. Water provides excellent shielding, if sufficient depth is maintained, and if no straight penetrations are created by the insertion of rods or air-filled tubes. Other reactors and critical facilities are shielded by massive walls usually of concrete and steel.

The surveyor *should* pay particular attention to locations where the biological shield is penetrated with holes. Holes may be required for insertion of control rods, entry and exit of coolant, external use of radiation beams, etc. Cracks may exist around plugs meant to close openings for insertion of fuels, insertion of samples for irradiation, etc. Iron has a much lower attenuation for neutrons near 25 keV than for other energies; iron rods or plugs in a reactor shield may permit the escape of significant numbers of neutrons of these energies. These potential sources of leakage *should* be surveyed frequently and remedial action taken when necessary.

The following specific requirements for shield surveys supplement the general requirements regarding frequency of surveys in Section 3.1.

(1) Comprehensive survey measurements *shall* be made during the first tests of a reactor to locate points of significant leakage through the shield. These surveys may indicate that alteration of the shielding is necessary before the tests may proceed safely.
(2) These survey measurements *shall* be repeated at power levels to be used in normal operation because the radiation field inside the reactor, the properties of coolants, etc., are different at different power levels and produce different amounts of radiation leakage through the shield.
(3) Surveys of this kind *shall* be repeated after any alteration of the reactor or its shielding or in the method of operation, or if personnel monitoring indicates an abnormal situation, and periodically throughout the life of the reactor.

3.3.7.3 *Residual Activity.* The operation of reactors and critical facilities results in the generation of fission products in the fuels and activation products in all materials exposed to the neutron flux. External dose rates of hundreds of rad per hour may be experienced from reactor fuels that have been used for long times in research reactors. Therefore, special procedures, equipment, and shielding are normally required for charging and discharging fuel and for handling irradiated fuel and samples. The radiation levels from the fission and activation products decrease rapidly in the first few hours after shutdown and cessation of fission. Personnel exposure *should* be reduced wherever possible by postponing entry into radioactive areas for a few hours after shutdown.

Activation products may be expected in the structure, shielding, coolant, and in anything placed in the neutron field. Parts of the reactor that may be removed, such as tubing, moderator, shielding, and irradiated samples will contain activation products that can produce serious contamination and exposure risks. It is important to recognize that some nuclides have very high neutron activation cross sections so that, even when they are present in small amounts not detectable by ordinary chemical analysis, their activation products may be the principal activity in a given irradiated material.

Water is a common coolant and moderator for reactors. The hydrogen and oxygen may be transformed to tritium and nitrogen-16. Except when heavy water is used, the amounts of tritium produced are normally very low. Neutron irradiation of boron and lithium will also produce tritium. The detection and measurement of tritium usually involves the collection of a sample and is difficult because of the very low energy of the beta rays emitted (E_{max} = 0.018 MeV). Instruments suitable for tritium measurements are discussed in Sections 5.2.2.1 and

5.2.3.5. Nitrogen-16 emits very high energy gamma rays (6.13 and 7.12 MeV) but fortunately also has a very short (7 s) half-life. It is therefore present only while the reactor is operating and then may contribute significantly to external personnel exposures. Impurities in the water also contribute to radioactivity in the cooling system. Build-up of neutron-activated corrosion products may be expected on all interior surfaces of the piping system, particularly where the flow characteristics change, such as at valves and couplings, and also in demineralizers, storage tanks, etc. For example, water and other coolants contain manganese-54 and cobalt-60 which may be formed by neutron activation of corrosion products of steel and stainless steel in the system. Fission products may appear from traces of fuel present on the surfaces of the fuel elements or from trace amounts of uranium in the water. It is recommended that surveys *should* include an evaluation of the trend of radiation levels which may indicate build-up of radioactive deposits within the facility.

Air is often used as a coolant for reactors and critical facilities and is also often trapped or dissolved in the water in the reactor. Nitrogen-16 is also produced from the oxygen in air. Activation of the noble gases in air gives principally argon-41, but also krypton-85, krypton-89, xenon-135, and xenon-137. These noble gases are principally a source of external exposure. Absorption through the lungs, and therefore internal exposure, is relatively unimportant.

Radioactive noble gases present some unusual surveying difficulties. They may mask the presence of more serious types of air contamination, for example that due to radioactive particles. Sometimes noble gases are mistaken for contamination on clothing; such apparent contamination will disappear very quickly when the clothing is in clean, circulating air. When exposed to these radioactive gases, ionization chambers and other instruments open to the air may continue to give erroneously high readings for some time after the air seems to be cleared because of gas still held within the instrument chamber; sealing such instruments in polyethylene bags helps to eliminate such gas collection. Methods and equipment for air sampling are discussed in Section 5.2.2.

In addition to anticipating the radiation fields and contamination resulting from removal of objects from a reactor, the surveyor *should* be alert for changes in the physical, chemical, and mechanical properties of materials due to irradiation. For example, capsules may develop leaks, they may explode or implode, solid materials may crumble to fine particles or dust, etc. Particulate contamination, radioactive noble gases, and radioiodines, especially iodine-131, may be released to the atmosphere and to the coolant system.

The cladding of irradiated fuel elements may become defective for a variety of reasons such as corrosion, over-heating, or weld failures. Leakage of fission products from any such defects can cause serious contamination of cooling or moderating systems or, if the defective element is outside the reactor, high level air or surface contamination.

The following recommendations are made with regard to monitoring the removal of irradiated materials from the core:

(1) Continuous monitoring of external radiation levels, air concentration, and personnel dose rate *should* be provided during removal and clean-up work. Particular care *should* be exercised while surveying the body surface of personnel to locate any small but intensely radioactive particles, that are capable of delivering high doses to the immediately adjacent tissues.

(2) All shielding for irradiated fuels and samples *shall* be surveyed in a manner similar to that described above for the reactor shield.

Measurements of radiation fields, and surface and airborne contamination around reactors, *should* be made as described in Sections 3.2.4 and 3.2.5.

4. Personnel Monitoring Methods

4.1 General

Measurements of the radiation exposure received by occupationally exposed individuals serve two different purposes. They provide information that may lead to the identification of undesirable practices and of unsuspected sources of high exposure, thus permitting the prompt application of controls to limit such exposure. They also provide some information regarding the exposure of the individual, permitting a comparison with long-term limits and guiding the establishment of any long-term controls that may be required.

The radiation surveys described in Section 3 may be used to predict the exposure of persons occupying the area surveyed. If the estimated doses are low and the exposure conditions predictable, the doses actually received may be assumed to be equal to these estimates. Thus the exposure of members of the general public in uncontrolled areas is estimated from the results of area survey measurements. If the estimated doses received by occupationally exposed individuals approach the maximun permissible dose, measurements made on or next to the exposed individuals are recommended. Because such measurements relate specifically to the individual they are a unique indicator of personal exposure during specific operations or over an extended period of time.

4.2 External Exposure Determination

4.2.1 *Requirements*

4.2.1.1 *Occupationally Exposed Persons.* Personnel monitoring for external exposure *shall* be performed on all occupationally exposed individuals who may receive more than ¼ of the applicable MPD during the normal course of their duties or through accidental exposure (NCRP Reports Nos. 33 and 39) (NCRP 1968a and 1971b).

If a field of mixed radiation is present in an area, for example a field of gamma rays and neutrons, it may be desirable to measure separately radiations that involve different critical organs or exhibit different quality factors. When the total dose equivalent from mixed radiation may exceed ¼ of the MPD, personnel monitoring *shall* be performed to measure any radiation that may contribute more than ⅒ of the MPD.

Personnel monitoring is unnecessary where the nature of the work performed or the nature of the radiation sources is such that personnel exposures are below the limits recommended for uncontrolled areas and where there is a very small potential for accidental exposure above these limits.

4.2.1.2 *Visitors.* Because of the diversity of radiation facilities and of visitors to their controlled areas, universally applicable recommendations for visitor monitoring are not feasible. Rules applicable in the particular circumstances of a given installation *should* be developed. As a general guide the following recommendations, consistent with those of NCRP Report No. 39, paragraphs 253–254 (NCRP, 1971b), are made.

Occasional visitors to controlled areas, including messengers, servicemen, and deliverymen, *should* be regarded as non-occupationally exposed persons. It is most improbable that such persons will receive in one year a dose equivalent exceeding the non-occupational limit of 0.5 rem during their brief occupancy of controlled areas. It is therefore unnecessary to provide personnel monitors.

Long-term visitors in an installation *should* be regarded as occupationally exposed if they are likely to receive a dose equivalent exceeding 0.5 rem per year and *should* be monitored according to the criterion of Section 4.2.1.1

4.2.1.3 *Accuracy.* For the purpose of monitoring the continued adequacy of radiation control methods in limiting personnel exposure, the measurement techniques require precision rather than accuracy (Section 5.1.1.1). Personnel monitoring procedures *should* be uniformly applied to different persons over a long period of time so as to permit comparisons between persons and periods. The desirable precision of measurements is ±10 percent.

There are two aspects of accuracy in personnel dosimetry if the objective is the estimation of critical organ doses: the accuracy of the instrument reading, and the accuracy of the assumptions used in translating the measurements into critical organ dose. The measurement accuracy desirable in personnel dosimetry depends on the radiation level to be measured. At the level of the MPD a measurement accuracy of ±30 percent *should* be achieved. If the dose equivalent to critical organs is less than ¼ of the MPD, personnel monitoring is not required and a lower level of accuracy (e.g., a factor of 2) is acceptable. On the other hand, at higher doses such as may occur during emergency procedures or accidents, determination with an accuracy better than ±20 percent is desirable.

The typical personnel dosimeter yields an acceptable approximation to the absorbed dose to the skin from commonly occurring radiations. For wide uniform beams the dose to deeper tissues may be deduced from measurements in phantoms for the particular type of radiation field. For penetrating radiations, the assumption commonly made that the dose to all critical organs is equal to the skin dose is conservative. However, this assumption may result, under other circumstances, in a large error in the estimation of dose to deep organs, particularly if the radiation is of low penetrating power or if the radiation field is highly localized. This error is of little importance at levels well below the MPD. When the measured skin doses are close to the MPD, careful consideration *should* be given to the point of measurement with respect to the critical organs and the location of the radiation field, and to the relation of skin dose to critical organ dose.

4.2.2 *Methods*

4.2.2.1 *Integration and Rate Systems.* Measurements of the external *exposure* of an individual may be obtained in either of two ways: by use of personnel dosimeters or by personnel exposure rate monitoring. In the first method, *exposure* or dose is integrated by a dosimeter carried on the person. In the second method, measurements of *exposure* rates and exposure times are made at positions very close to personnel during the course of work in a radiation field. The choice of method will depend on the type of radiation, the *exposure* rate, its time-variation, and the need for immediate read-out.

Measurements made by means of personnel dosimeters have the following advantages over measurements obtained by exposure rate monitoring (Section 4.2.2.6):

(1) the dosimeter provides an automatically integrated reading over the period the dosimeter is carried;
(2) the measurement is made without employing and exposing additional personnel for making the measurement;
(3) a large number of people can be monitored in a uniform fashion;

(4) exposures received accidentally or from unknown sources will be included in the measurement, provided the range of the dosimeter is not exceeded.

On the other hand, in comparison with exposure rate monitoring, dosimeters have the following disadvantages:

(1) they may not respond to all types of radiation encountered;
(2) they provide a measurement at a single preselected point on the body, which may not be the point of maximum dose;
(3) they provide information on radiation fields after the exposure has occurred and often not rapidly enough to permit short-term remedial action. Dosimeters of the signaling type or ionization chambers of the self-reading type may be read with sufficient rapidity to permit such control if radiation conditions are not varying rapidly.

4.2.2.2 *Estimation of Whole-Body Dose.* If it is likely that the body may be exposed fairly uniformly, a dosimeter *should* be worn on the trunk of the body since the gonads and most of the blood-forming organs that constitute the principal critical organs for whole-body exposure are located in the trunk. Suitable locations are the breast pockets, lapels, and the belt. Dosimeters *should* be worn so that they are visible at all times, except when they are intentionally covered by a shield. This prevents unintentional shielding by clothing or by items in a pocket.

When the trunk of the body is largely shielded by protective clothing (e.g., a lead-rubber apron), it would be improper to wear a single dosimeter on the outside of such clothing since doses to the whole-body, the gonads, and most of the red bone marrow would then be greatly overestimated. Measurements (e.g., Buchan, 1968; Bushong *et al.,* 1969; Webster, 1969; Wold *et al.*, 1971) have shown that, when a lead-rubber apron is worn by radiological personnel conducting fluoroscopic examinations and catheterization procedures, the *exposure* of the face and neck will exceed the *exposure* recorded under the apron by factors between 6 and 25. Under these circumstances the thyroid gland and the lens of the eye will become the critical organs, and their *exposure* should be monitored. Ideally, a second dosimeter *should* be worn at the collar level for this purpose (Section 4.2.2.3). If only one dosimeter is worn and one of its purposes is the estimation of "whole-body" dose, it is recommended that it be worn on the trunk under the apron. The Radiation Safety Officer *should* in this event determine, through phantom or personnel measurements, average factors by which the recorded dose *should* be increased to express thyroid and/or lens doses. If, as has been proposed (Bushong *et al.,* 1969), a single badge is worn on an unshielded part of the trunk, such as at the collar, then this fact *should* be recorded and the measurement used to estimate the thyroid or lens dose.

If the radiation field is largely due to external beta radiation, separate determinations may be required of the dose to the skin and to deeper critical organs. This separation is normally achieved by using suitable absorbers on the dosimeter or dosimeters. In a mixed field of gamma rays and neutrons, separate dosimeters or a complex dosimeter having parts with different, but known sensitivities to the radiations *should* be worn (See Section 5.1.2.4).

4.2.2.3 *Partial Body Exposure.* The requirements for wearing dosimeters may make it desirable or necessary to wear dosimeters on more than one part of the body. For example, in a non-uniform radiation field in which both the hands and the whole body may receive significant fractions of the applicable MPD, dosimeters *should* be worn on both the hands and the trunk of the body. Similarly, as discussed in Section 4.2.2.2, when the trunk of the body is shielded, a second dosimeter *should* be worn to monitor the exposure of the head and neck (Langmead and Farmer, 1971).

Where sealed or unsealed radioactive sources are handled, it may be particularly important to determine the dose to the hands. Extremity dosimeters *should* be worn as near to the point of maximum exposure as possible (on a finger or the wrist) and *should not* be shielded from the radiation by the extremity. A wrist dosimeter *should* be worn on the wrist that the wearer estimates will receive the greater dose and usually on the inside of the wrist. It may be desirable to apply a finger-to-wrist dose ratio in evaluating wrist dosimeter readings (Chiswell and Gilboy, 1972).

Where work is observed through viewing ports in shielding, a separate evaluation of the eye dose may be required. Practical dosimeters for measuring the dose to the lens of the eye are not available. When the radiation field is large enough, the required measurements can be obtained from a dosimeter affixed to a head-covering above the eyes. If this is not feasible, measurements obtained by personnel exposure rate monitoring are advisable.

4.2.2.4 *Choice of Dosimeters.* For x rays, gamma rays, and electrons the choice at present lies between ionization chambers, film badges, photoluminescent glasses, and thermoluminescent dosimeters (Section 5.1.2.1). The principal advantages of ionization chambers, particularly those of the self-reading type, are the simplicity and the speed with which readings are made. They are therefore particularly useful for monitoring exposures during non-routine operations or during transient conditions, or for monitoring short-term visitors to an installation. Chambers *should* be tested for leakage periodically and those that leak more than a few percent of full-scale over the period that they would be used *should* be removed from service. The other available systems have two main advantages over ionization chambers:

(1) their capability for use with filter systems that provide an approximate energy discrimination and permit separate estimates of dose from high-energy radiation, which irradiates deeply-seated critical organs, and soft radiation, which irradiates only the skin or eyes; and

(2) their freedom from leakage that permits long periods for the integration of exposure.

However, all these devices are subject to some "fading" of the effect produced by radiation and *should* be read promptly at the end of the integrating period (Becker, 1973).

For neutron fields the practical devices are nuclear-track film, thermoluminescent dosimeters containing ^{6}LiF, and fission-track-counting systems (Section 5.1.2.4). The nuclear-track films do not respond to neutrons below about 0.5 MeV in energy. In practice a substantial fraction of the neutrons may be below this energy. Track-counting is a relatively insensitive method of neutron dosimetry. For low doses, counting of a statistically significant number of tracks is too time consuming to be warranted. On the other hand, at high doses it is difficult to distinguish tracks from one another so that they can be counted. Fading occurs and, as a result, short tracks may disappear. For these reasons nuclear-track film is more useful in demonstrating that large neutron doses have not been received than in measuring actual doses, particularly if they are less than the MPD.

The ^{6}LiF and fission-track-counting systems do not suffer from these disadvantages and will provide measurements at or below MPD levels. These methods are sensitive down to doses of a few millirad and down to thermal neutron energies. Personnel exposure rate monitoring is often employed to obtain measurements of neutron dose at rates lower than those corresponding to the MPD.

4.2.2.5 *Integration Period.* The time over which exposure should be integrated with personnel dosimeters depends on the average reading expected and on the characteristics of the dosimeters, particularly their sensitivity, reproducibility, and leakage (or fading) rate. At the lowest levels of whole-body occupational exposure for which dosimeters are advised, a monthly integrating period is recommended for film badges and silver-activated phosphate glasses, and a weekly integrating period for pocket chambers, in order to secure reasonable precision. Both the LiF and the CaF_2:Mn thermoluminescent systems are sensitive enough to be used over a period of a week or less, while their stability is sufficient to permit use over monthly or quarterly periods.

If whole-body doses of the order of one-half of the MPD are expected, integration over one week or two weeks is desirable for film badges, TLD systems, and glasses; and over one week for pocket chambers, to permit closer control of individual exposures and to reduce errors due to fading and leakage.

4.2.2.6 *Exposure Rate Monitoring.* Where radiation fields may change rapidly and unexpectedly, and where rapid control of radiation exposure is desired during a particular operation, personnel exposure rate monitoring *should* be performed. Measurements of *exposure* rate and exposure times are made continuously at positions very close to personnel, and integrated *exposures* are computed from these measurements. This technique *should* also be adopted when dosimeters sufficiently sensitive to the radiation involved are not available. Local doses arising from the deposition on skin of highly radioactive particles *should* also be evaluated by rate monitoring methods.

Personnel exposure rate monitoring *should* be performed by well-trained persons familiar with the nature of the work being monitored, the magnitude and origin of possible high rates of exposure, and the methods required for rapid control of unwarranted exposure.

4.2.3 *Interpretation*

Usually monitoring instruments do not measure directly in dose-equivalent units (rem), although there are some instruments that, for certain types of radiation and for certain organs, are calibrated directly in rem. The measured quantities include *exposure*, fluence, kerma, optical density, current, volts, and counts per minute. The absorbed dose at the point of measurement is usually derived from the measured quantity by the application of a calibration factor (Sections 5.1.1.1 and 5.1.2). The dose equivalent is derived from the absorbed dose through the application of appropriate physical and biological factors (see Section 2.1.1).

Unless the body is subjected to a uniform distribution of dose, the "whole-body dose" and doses to critical organs cannot be strictly determined from measurements at one point or a few points. When personnel doses are well below the MPD, it may be assumed for personnel monitoring purposes that the surface dose determined at one point on the trunk of the body or on part of the body, such as the hand, represents the dose to the whole body or to that part. Compliance with codes controlling maximum permissible doses may be demonstrated on the basis of surface doses. At levels approaching or exceeding the maximum permissible dose, the dose to the whole body and the critical organs *should* be more carefully evaluated and correction factors relating to the circumstances of the exposure *should* be applied. These *should* include the effects of the exposure geometry and radiation energy (Section 4.2.1.3).

4.3 Internal Exposure Monitoring

4.3.1 *Requirements*

4.3.1.1 *Need.* There are two objectives in undertaking measurements of internal contamination of persons employed in areas in which radioactive materials are used, handled, or stored. The principal objective is to indicate qualitatively whether entry of radionuclides into the body has occurred. The second objective is to determine the organ or body burden of radionuclides and to estimate the resultant internal doses. Internal monitoring procedures serve as an adjunct to external contamination surveys. For radionuclides such as tritium, which require special survey instrumentation, periodic bioassay may be used as a principal method of monitoring.

The routine determination of internal contamination is usually necessary only in installations where unsealed radioactive material may become airborne and where air concentration levels may lead to depositions exceeding 10 percent of the maximum permissible body (organ) burdens. If there is evidence of contamination during external surveys, the likelihood of significant intake of radioactivity during routine handling procedures will be increased and internal contamination monitoring *should* be seriously considered. For example, if measurements of airborne contamination in a work area normally show levels exceeding 10 percent of the maximum permissible concentration, a program of bioassay or whole-body counting *should* be provided for personnel in that area. Such a program *should* also be instituted if removable radioactive contamination by radionuclides of long effective half-life is frequently observed in working areas in amounts of the order of the maximum permissible body burden, in locations and under circumstances where appreciable intake is likely.

In practice, routine monitoring for internally deposited sources is utilized principally in the following types of operation:

(1) those in which particulate, gaseous, or volatile radioactive materials are handled in large quantities in unsealed form, e.g., radioactive iodine in iodination procedures involving 1 mCi or

more; and tritium and its compounds in tritium labeling by exchange reactions, in the luminizing industry, in heavy water reactors, and in commercial production plants

(2) facilities for the mining, milling, refining, or fabrication of uranium, e.g., as reactor fuel elements; and

(3) facilities for the processing of plutonium and other transuranic elements.

If intake of radioactive material is suspected, such that the average body burden over a three-month period may approach the maximum permissible body burden, bioassay of an appropriate kind or whole-body counting *shall* be carried out promptly. Such action *should* be taken at lower levels of potential contamination where radionuclides of long effective half-life (e.g., plutonium-239) are involved. Estimates of internal dose derived from these measurements can be helpful in deciding on appropriate medical measures. If treatment is instituted to reduce the body burden, follow-up measurements *should* be made to permit an evaluation of its effectiveness.

4.3.1.2 *Frequency of Measurements.* The frequency of routine measurements depends principally on the effective half-life of the radioactive contaminant, the variation of excretion rate with time, and the recent experience of the individual and the group with respect to internal contamination levels. Tritium oxide, for example, has a biological (effective) half-life of about 10 days and therefore routine urine samples obtained every two weeks are appropriate for persons who may have body burdens up to the maximum permissible level (23 μCi/liter of urine) (Healy, 1970). For persons suspected of having larger intakes, the sampling frequency *should* be increased to facilitate exposure control. For materials of long effective half-life, such as plutonium, the period between measurements *should* be chosen to allow monitoring of the long-term build-up; quarterly measurements *should* be adequate for persons with the highest potential intake and annual measurements for those with minimal but real exposure potential.

4.3.1.3 *Methods.* The determination of the dose to the body or to specific organs due to internally deposited radioactive materials cannot be made directly. There are, however, three indirect methods:

(1) In the first method, known as bioassay, the body burden is estimated by measuring the radioactivity in collections of urine, blood, breath, or feces and relating the excretion rate to body burden by the use of biological models.

(2) In the second method, known as *in vivo* or whole-body counting, an estimate of the body burden of gamma-emitting nuclides is obtained by counting the gamma rays emitted from the body

and analyzing the pulse-height spectrum. This technique can also be used to measure the *bremsstrahlung* from energetic beta emitters.

(3) The third method derives from survey data, particularly air concentration levels, which permit estimates of intake of radioactive material, followed by estimates of the fraction reaching individual organs. The estimated organ burden must be modified by applying known data on excretion or organ turnover.

Radiochemical analysis of bioassay samples and whole-body counting are highly specialized procedures for which most users of small quantities of radioactive materials, particularly alpha emitters, are not equipped or staffed. These services are available from commercial companies. Many universities and national laboratories are now equipped with whole-body counters (IAEA, 1970a).

4.3.1.4 *Accuracy.* For the purposes of radiation protection, the desirable accuracy of activity or dose estimates *should* be within ±30 percent, particularly at levels of the order of the maximum permissible dose.

In practice, the accuracy with which activity is estimated through whole-body counting is often better than ±30 percent, and it is, therefore, the method of choice for the radionuclides that allow its use. By contrast, the process of translating bioassay measurements and intake estimates into body or organ burdens has inherent inaccuracies that arise from the uncertainties in the biological models employed, and from variations between individuals (Sill *et al.*, 1964). Thus, organ or whole-body burdens are usually estimated from bioassay and survey data with errors considerably greater than a factor of 2.

4.3.2 *Bioassay Procedures*

Bioassay is the analysis of body fluids, excreta, or tissue samples (such as hair) for their concentrations of radioactivity. The type of sample most appropriate for analysis will vary with the contaminating element, its physical and chemical form, and its mode of entry to the body (Jackson and Dolphin, 1966). This information may be translated, by the application of biological factors, into estimates of body or organ burden. The method has principal application to those radioactive substances that are soluble (since only relatively soluble substances will be rapidly transferred to body fluids, such as blood and urine), whether inhaled or ingested. Inhaled material that is insoluble in body fluids is likely to be retained in the lungs and not detected by assay of blood or urine. As previously mentioned, however, inhaled particulates will in part be swallowed and will, especially if insoluble, appear in the feces which may be assayed.

An important advantage of bioassay over whole-body counting is the ability to detect alpha- and beta-emitters readily, and to identify them. Another advantage is that samples can be collected and measured without requiring the attendance of the potentially exposed person at the measurement facility. Disadvantages include the necessity for subject cooperation in sample collection and the delayed availability of results due to the time needed to perform the chemical and analytical procedures. For meaningful results, care must be taken to obtain representative samples that are not contaminated by the collecting procedure.

Analytical procedures for many common radionuclides are described in the literature, particularly for tritium, radium, thorium, uranium, plutonium, polonium, iodine, cesium, phosphorus, strontium, barium, and ruthenium (WHO, 1966; DHEW, 1965; DHEW, 1967). NCRP Report No. 30 (NCRP, 1964) contains general information on bioassay procedures. Detailed discussions of the relationship between radionuclide intake and concentration in excreta at times subsequent to intake are contained in ICRP Publications 10 and 10A (ICRP, 1968a; ICRP, 1968b).

4.3.2.1 *Urinalysis.* The analysis of urine samples is the most common method used for routine estimation of body content of radionuclides. The analysis is usually specific for a radioelement or nuclide. Analysis for total alpha or beta activity may serve to indicate that no deposition exists, precluding the need for performing specific analyses. Before an analysis for total beta activity is performed, the natural ^{40}K content of the urine must be removed, otherwise it will mask trace amounts of other radionuclides. The methods employed *should* be sufficiently sensitive to permit the measurement of sample activities corresponding to those excreted by individuals containing small fractions of the maximum permissible body burden. The activities of interest may be of the order of 0.1 disintegration per minute in a 24-hour collection[14] for long-lived alpha-emitters, and a few nanocuries for the most radiotoxic beta-ray or gamma-ray emitters. Methods for such measurements are discussed in Section 5.2.3.

The body burden is estimated by employing mathematical models describing the metabolism of the radionuclide and particularly the excretion rate as a function of time after the termination of exposure (ICRP, 1968a; ICRP, 1968b). These models are based on metabolic studies in humans and other mammals. For a few nuclides, such as tritium or the sodium isotopes, that are loosely bound in the body, the

[14] For example, for plutonium, 14 dpm in the urine excreted in the 24-hour period approximately 30 days after an acute exposure indicates internal deposition of approximately one maximum permissible body burden (0.04 μCi).

excretion rate can be described by a single exponential for long periods of time so that the ratio of excretion rate to body burden is constant. For example, for tritiated water an equilibrium is rapidly established between the concentration of tritium in body fluids and in urine, and a maintained urine assay of 23 μCi/liter signifies that a maximum permissible body burden is present (Healy, 1970). Generally, however, excretion patterns are complex; many may be described by a combination of exponentials reflecting several compartments (ICRP, 1968a). In ICRP Publication 10, Appendix C, the excretion functions of 30 common radionuclides are described, and of these 7 appear to follow simple exponentials and 12 follow complex exponentials, while 11, principally the isotopes of the alkaline earths (Ca, Ba, Sr, and Ra which are bone seekers), appear to follow power functions of time (Sanders, 1960). The excretion function is the time derivative of the retention function, and therefore knowledge of the excretion can be related to the body burden. Usually knowledge of the time of exposure and the analysis of a series of samples are required before a valid estimate of the magnitude of the initial body burden can be made. The estimation of body burdens is facilitated by the development of computer codes for routine use in larger bioassay facilities.

The accuracy of body burden estimates based on urinalysis is greatly affected by several variables. Generally, subjects do not submit 24-hour urine samples but rather a limited sample which is an unknown fraction of the daily output. Estimates of this fraction may be derived by measurement of the creatinine content, the daily output of which is relatively constant (Jackson, 1966). There is considerable variability (of the order ± 25 percent of the mean) of the renal clearance rate between normal individuals (Pochin, 1964) and of urine output (ICRP, 1975) which affect the fraction of the body burden excreted per day (Beach and Dolphin, 1964). Even for a given individual this fraction will vary from day to day (Müller *et al.*, 1961).

The principal usefulness of urinalysis in a radiation protection program is as a qualitative indicator of intake of radioactive materials and, therefore, of working conditions requiring surveillance and possible correction. Because of its inherent inaccuracies and limitations, the technique affords only a semi-quantitative index of body burden, with the notable exception of tritium.

4.3.2.2 *Blood Analysis.* The radioactive content of the blood provides information similar in kind to that provided by urinalysis for total body burdens. Since the collection of a blood sample lends itself less to routine procedure than the collection of urine, it is rarely used as a bioassay technique, except after suspected acute intake of soluble radioactive material. The blood sample *shall not* be drawn through contaminated skin. Blood analysis is of considerable importance as a method of determining acute neutron exposure by means of radioassay of the neutron-activation product sodium-24 (Harris, 1961). This method is discussed in Section 6.3.1.

4.3.2.3 *Analysis of Sputum, Nasal Smears, and Feces.* Bioassays of sputum, nasal smears, throat swabs, and feces are normally used to determine whether radioactive material has been inhaled or ingested. Some of the inhaled airborne particulates, especially those of large particle size (mean aerodynamic diameter greater than 10 μm), will be deposited in the nasal and upper respiratory passages, transferred to the throat by ciliary action, and subsequently swallowed. Much of this material, particularly if insoluble, will appear in the feces. These assays are of particular value when surveys suggest that intake of insoluble radioactive materials may have occurred during an accident or through a previously unsuspected chronic contamination, and they are useful in providing qualitative confirmatory information.

4.3.2.4 *Breath Analysis.* For specific radionuclides that may exist in gaseous compounds (e.g., $^{14}CO_2$) or that produce gaseous daughter products, breath analysis may provide quantitative information on body burden. A method of determining the whole-body burden of radium-226 is the analysis of exhaled air for radon-222, its immediate daughter product, about 70 percent of which is eliminated from the body on the average (Vennart *et al.*, 1964). It has also been shown that the concentration of tritium in exhaled water vapor is essentially the same as that in urine. The concentration of tritium in the condensed vapor may be measured in a liquid scintillation counter with a sensitivity of about 1 μCi/liter of water (Chiswell and Dancer, 1969). This method may be preferable to the more usual urine analysis for tritium.

4.3.3 *Whole-Body and Other External Counting*

External counting of the entire body or a specific organ is the method of choice for estimating the burden of gamma emitting and some beta emitting radionuclides. This method, however, requires special and expensive equipment with high sensitivity and low background such as that described in Section 5.2.4 (Marinelli *et al.*, 1962; Rundo, 1962). When the body burden of a gamma-ray emitter is roughly 1 μCi, instruments, with little or no shielding, and less sensitive than a whole-body counter may be used for its measurement in the body.

Whole-body counters are employed for two purposes in radiation protection programs. They are used principally to measure the body burdens in a sample of the employees at an installation at various times. If the average body burdens are well below the MPBBs, radiation protection control is judged to be adequate. Members of the population living near an installation may be similarly studied. The other purpose is to study persons who have been exposed to possible internal deposition of radionuclides in an accident.

Special care must be taken to insure that there is no external contamination on the clothing or body of the subject, not only because spurious measurements will be obtained, but also because the facility may become contaminated. Showering and the wearing of special clean clothing is recommended.

Interpretation of the count-rate from the body in terms of body burden requires that the counter be calibrated for the radionuclides involved. One method is to distribute known sources in phantoms in a geometrical arrangement that simulates the body position. Counters have also been calibrated by means of measurements on persons with known body burdens. Furthermore, assumptions about the distribution of radioactive materials in the body and about the absorption of radiation by the body are necessary. Emitters of energetic gamma rays, that are minimally absorbed by the body, are measured with the greatest accuracy. Furthermore, the use of pulse height analysis for a specific nuclide through the measurement of its photopeak(s) has greater accuracy for higher energy gamma rays. The method is, therefore, particularly suited for determinations of many nuclides including iron-59, cobalt-60, zinc-65, ruthenium-106, iodine-131, cesium-137, and radium-226. (For radium-226 determinations, breath measurements may also be required if the time interval between exposure and measurement is less than one year.)

External measurements may also be made on parts of the body to evaluate the radioactivity in specific organs. This method is particularly appropriate for the measurement of radioiodine in the thyroid gland because of the superficial location of the thyroid and the high concentration of iodine by the gland. The count-rate of one or two sodium-iodide scintillation counters in contact with the skin directly over the thyroid is compared with that produced by a known standard in the same geometry. In order to make the method sensitive enough to measure nanocurie levels, crystals with a large cross-section are necessary, and the measurements *should* be made inside a shielded enclosure, such as a whole-body counter room, to keep the background low (Laurer and Eisenbud, 1963; Wellman *et al.*, 1967; Blum and Liuzzi, 1967). Both G-M and small scintillation counters have been used to measure sodium-24 produced by neutron activation of the body sodium in people exposed to neutrons in criticality accidents. The counter is placed over the back or the abdomen. Care must be taken to assure that external contamination or radiation from external sources does not interfere with the measurement (see Section 6.3.1).

Scintillation counters with NaI(Tl) detectors 2–5 inches (5–12 cm) in diameter have been used to measure cesium-137 in people. The counter is placed in the person's lap and he bends forward over it while being counted. If careful corrections are made for background and potassium-40 content by counting a person of similar size but containing little or no cesium-137, it is possible to measure the body burden nearly as accurately as with a whole-body counter. The x rays from ^{239}Pu can also be measured with large area external detectors (see Section 5.2.4.4).

4.3.4 *Estimation of Intake and Organ Burden from Air Concentrations*

The advantage of estimates of intake and organ burden from air concentrations is that they require no special procedures to be performed on a person. No special instrumentation beyond that used in survey procedures is needed, no biological samples are required, and no radiochemical procedures are involved. This method is the only one readily available for many small users of radioactivity, and may be the only one feasible for estimation of body burdens of radionuclides with very short half-life. Nevertheless, if the body burden is estimated to be significant, efforts *should* be made to confirm the estimate by bioassay or whole-body counting.

The basic data consist of measurements of air concentrations from which the average concentration may be calculated (Section 3.2.6.1). The body burden is assumed to be proportional to the average air concentration and to the intake periods for times which are short compared with the biological half-life.

The disadvantage of the method lies in its inaccuracies and in the need to make biological assumptions that, at best, can apply only to the standard man. Uncertainties regarding intake arise because the concentration at the breathing zone may differ from that of the air volume actually sampled and may vary with time, and because the volume of air inspired is not precisely known. Moreover, the fraction of airborne activity retained in the lungs or entering the gastrointestinal tract, and the fraction that enters the blood, are heavily dependent on the physical and chemical properties of the active material, particularly its solubility and its particle size. Larger particles are deposited in the upper respiratory tract and usually find their way into the gastrointestinal tract through swallowing. Small particles tend to be retained in the lungs, whence, if soluble, the material passes into the blood stream. Hence, no one set of biological factors is appropriate for universal use in converting air concentration into body or organ burden (Morrow *et al.*, 1967; ICRP, 1966).

4.4 Personnel Contamination Monitoring

The method and instruments used for personnel contamination monitoring do not differ appreciably from those used for other types of contamination surveys. Personnel surveys *should* be made occasionally during work with radioactive material. When persons leave a controlled area, a more thorough survey *should* be made, particularly of the hands, and, if spillage is suspected, of the shoes.

4.5 Investigation of Unusual Exposure

4.5.1 *General*

Regardless of precautions that may be taken to regulate working conditions and to control personnel exposures, situations will occur that require investigation to determine the extent of the exposure. Most of these situations will be readily apparent; however, some may be only subtle hints that an unusual condition or situation exists or has occurred. Emergency situations are not included in this section since they are discussed in Section 6 of this Report.

4.5.2 *External Exposure Investigation*

4.5.2.1 *Situations Requiring Investigation.* Typical circumstances that may require investigation to resolve questions regarding external exposures are:

(1) The loss of a dosimeter or of a dosimeter result.
(2) An abnormal dosimeter result. An investigation may include the possibilities of artifacts, non-occupational exposure, equipment failure, breach of work rules, and radioactive contamination.
(3) Dosimeter results that may not be a valid measure of the exposure received; for example, a dosimeter was not exposed to

the same conditions as the major portion of the person's body, or did not record an expected significant localized exposure.

(4) Loss of exposure control resulting in uncertain or unknown exposures.

4.5.2.2 *Prompt Investigative Action Associated with External Exposures.* Investigation and evaluation of likely personnel exposure may require further action as follows:

(1) Immediate processing of dosimeter.
(2) Measurement of dose rate and estimation of exposure duration. If such data have not been previously obtained, it may be necessary to simulate the exposure conditions.
(3) Collection of supporting dose measurements primarily for exposure control purposes.
(4) Determination of type of radiation (beta, gamma, x ray, neutron, etc.) and approximate energies if known.
(5) Collection of source size data. This may be the size of a beam or particulate contamination on skin or clothing.
(6) Autoradiographic examination of particulate contamination. This should be interpreted to reflect the dose rate to 1 cm^2 of tissue.
(7) Determination of previous external exposure history for the individual to ascertain if any controls or limits may have been exceeded.

4.5.3 *Internal Exposure Investigation*

4.5.3.1 *Situations Requiring Investigation.* Typical of circumstances that may require investigation to resolve questions regarding internal exposure are:

(1) Airborne concentration well in excess of established limits.
(2) Skin or clothing contamination exceeding a significant fraction of body burden or MPD for skin.
(3) Spillage or leakage of radioactivity; for example, from a ventilated hood or glove box to the floor or work area.
(4) Potentially contaminated wounds. This would include absorption either through intact but damaged skin or through breaks in the skin caused by a cut or puncture wound.
(5) Unexpected results from a routine surveillance program such as bioassay sampling or *in vivo* examination.
(6) Release of radioactivity resulting from failure of containment system.

4.5.3.2 *Prompt Investigative Action Associated with Internal Ex-*

posures. If an intake of radionuclide is suspected, immediate action such as decontamination is necessary to prevent additional exposure. In addition, prompt investigative action is necessary to evaluate the seriousness of the intake and to determine future corrective action. Investigative action may include:

(1) Retrieval of sample collector from any air samplers used in the location and the measurement of the activity present.

(2) Measurement of airborne concentration, if still elevated, at a location representative of the breathing zone.

(3) Estimation of probable duration of exposure which, together with airborne concentration and nasal smear activity level, may be the only immediately available indicators of the probable deposition.

(4) Measurement of contamination in nasal smears, sputum samples, nose blows, etc. Alpha activity greater than 40 dpm or beta-gamma activity greater than 100 dpm, measured in a well counter, may indicate possible internal deposition. (See forthcoming NCRP Report on management of persons contaminated with radionuclides".)

(5) Scheduling collection of bioassay samples including urine, feces, and/or blood as appropriate.

(6) *In vivo* examinations including whole-body counts and/or counts of selective body areas such as thyroid or wounds.

(7) In connection with wounds involving radioactive material, examination of the wound and any excised tissue to assure that all or most of the contaminant has been removed.

(8) Identification of the radionuclide(s) involved to assist in evaluation of internal exposure.

(9) Determination of the chemical form of the radionuclide(s) and its solubility in body fluids to assist in later evaluation of internal exposure.

(10) Determination of the physical characteristics of the material including particle size measurements of specimens collected from air samples, nasal smears, skin, or clothing. These data are useful in evaluation of internal exposure by means of the ICRP Lung Model (ICRP, 1966).

4.5.4 *Follow-Up Action*

Based on data collected or action taken during the investigation, several courses of action must be considered. These include:

(1) A preliminary evaluation to determine if medical treatment,

such as surgical removal of contamination, and/or administration of chelating agents, is necessary; and/or temporary work restriction is warranted until more data are available for better evaluation of the exposure. Generally, these decisions will have to be made with only a minimum of data available. If prompt action is not taken within a few hours, much of the benefit from treatment may be lost.

(2) A search for evidence that other personnel inside, or persons outside the installation, may have also been exposed as a result of the incident.

(3) A final evaluation of the incident that summarizes the exposure of persons involved and recommends corrective action to minimize the likelihood of similar incidents in the future.

4.6 Organ Dose Equivalent

Only rarely is it necessary to compute the organ dose equivalent for internal emitters. An example would be ingestion or inhalation of a radionuclide resulting in a body burden greater than the maximum permissible level, which may require restriction of exposure to external radiation fields and/or internal emitters (ICRP, 1964; ICRP, 1968a). Such a computation is difficult because the activity in an organ and therefore the dose rate will usually vary as a function of time, dependent on the rate of intake of radioactive material, its metabolic properties, and physical half-life. It is therefore necessary to make assumptions regarding internal activity concentrations between the sampling dates. It can be assumed, for example, that the intake producing the organ burden occurred immediately after the previous sampling date. ICRP Publication 10 (ICRP, 1968a) provides further discussion of this problem and guidance on the calculation of organ dose equivalent. Because of this difficulty common practice is to make periodic comparisons between estimated body burdens and the maximum permissible body burdens recommended in NCRP Report No. 22 (NCRP, 1959).[15] The MPBBs have been computed from weekly dose equivalent limits for the critical organs or by comparison with the MPBB for radium-226.

In the context of a high internal dose, when the organ dose equiva-

[15] The maximum permissible body burden for radioisotopes of iodine given in NCRP Report No. 22 should be reduced by a factor of 2 to reflect the revision of MPD for the thyroid gland in NCRP Report No. 39 (NCRP, 1971b).

lent from external exposure is greater than 25 percent of the MPD, it *should* be added to the internal dose equivalent. If the total organ dose equivalent approaches or exceeds the MPD, accurate estimates of the external component *should* be made that include corrections for depth dose, shielding, and geometrical factors.

4.7 Personnel Radiation Exposure Records

4.7.1 *Requirements*

Since maximum permissible doses are set for long periods of time, while personnel exposure data are gathered over relatively short periods such as weeks or months, it is necessary to record and retain each measurement made. Maintenance of such records for personnel who may receive significant fractions of the Maximum Permissible Dose is mandatory under most protection codes. In addition to satisfying the legal requirement of compliance with regulations, the records provide the basis for evaluation of a radiation control program and for initiating changes in such a program. They may also be of value in medico-legal disputes.

Personnel exposure records *should* include the following categories:

(1) Records of external exposure
(2) Records of internal exposure
(3) Records of unusual exposure
(4) Records of exposure from previous employment
(5) Records of special investigations.

Exposure histories are necessary for determining the cumulative exposure of individuals and useful for evaluating the significance and origin of internal exposure when detected.

Further guidance on the structure and content of records may be obtained from ANSI (1966).

4.7.1.1 *Scope of Records.* Complete records are particularly important for individuals who may be exposed to high dose levels. Unusual exposures *should* be carefully documented since they may be due to failure of control measures or may indicate the desirability of new control measures.

All personnel exposure records *should* identify:

(1) the individual, including name, date of birth, Social Security number, work location, and previous radiation work;
(2) the employer;

(3) the purveyor of dosimetry services; and
(4) the person responsible for radiation protection and his title.

Records of external exposure *should* include:

(5) the period during which a dosimeter was carried or personnel monitoring was performed;
(6) the critical organ (such as skin, extremities, lens of eye) to which the particular data apply;
(7) for unusual exposures, the type and energy of the radiation that produced the exposure;
(8) the computed dose equivalent for the specified period and the quality factors assumed;
(9) the summation of dose equivalent for the whole body (and for other critical organs if the levels approach the MPD), for the periods over which Maximum Permissible Doses are set (per quarter, per year, or lifetime).

Records of internal exposure *should* include:

(10) the type of radiation (e.g., alpha, beta);
(11) the identity of the radionuclide(s), and chemical and physical form, if known or suspected;
(12) the collection period for and date of the sample;
(13) the type and size of sample;
(14) the concentration of activity in the sample;
(15) results of organ and/or whole body counting;
(16) for unusual exposure, the estimated body burden, along with any information on distribution in the body;
(17) for large unusual exposure and when there is also significant external exposure, the computed dose-equivalent for a specified period;
(18) reference to the method of estimating body burden and/or computed dose equivalent.

Records *should* be maintained for exposures of the parts of the body for which different MPDs apply (e.g., whole body, skin, extremities, specific organs). Unless there is reason to believe that external dose estimates apply only to particular parts of the body, it *should* be assumed that they apply to the whole body.

Where both external and internal exposures approach the MPD, the total dose-equivalent received by the whole body or specific organs and the quality factors used in its computation *should* be recorded where possible for the periods over which the MPDs are defined.

Exposures determined by personnel exposure rate monitoring and not already included in the personnel dosimeter record *should* be included in the exposure record of the individual involved. Such exposures may have been excluded from the personnel dosimeter record because personnel dosimeters were carried at unsuitable locations, or were incapable of measuring the type of radiation involved, or were not carried. On the other hand, when a dosimeter is carried for monitoring a specific operation and its reading duplicates part of the long-term reading of another dosimeter, then only the latter reading *should* be entered into the cumulative record. Separate operational records *should* be maintained of such short-term readings to aid in the safety evaluation of specific operations.

4.7.1.2 *Supporting Records.* A complete record of personnel exposures *should* include the original data in the form in which they were measured (e.g., density, counts per minute, neutron-induced tracks per area observed), control or background readings, size of aliquot measured, and calibration data. Such records *should* be maintained at the installation or by the purveyor of the protection services.

The record system *should* contain all policies and standards for control of personnel exposure, together with a description of the methods by which the measurements and evaluations were made, and an estimate of the likely probable error. The records *should* also include the identity and calibration factors of counting equipment, monitoring instruments and dosimeters, and data on the dates and method of calibration or a reference to where this information can be found. Biological data used to relate excretion rate to body burden *should* also be recorded.

4.7.2 *Retention of Records*

Because of the long latent period of some radiation-induced diseases, the records specified in Section 4.7.1 *should* be retained for at least thirty years.

Employers who receive radiation protection dosimetry services from an outside supplier *should* arrange with him to retain the supporting data for each individual measurement for the required period.

Annual summaries, records of unusual exposure, and supplementary information on the design and execution of the radiation protection program *should* be retained until it is no longer necessary to provide exposure histories. The actual retention time *should* be determined by reference to applicable codes.

4.8 Evaluation of Personnel Monitoring Data

Personnel exposures *should* be examined regularly by the individual responsible for radiation safety in the installation. They *should* be compared with the levels established by past experience as normal for a particular job and location.

Critical examination of both long-term and short-term personnel doses may reveal where modifications of facilities or procedures may be worthwhile or required for the reduction of exposure. Recommendations for modifications *should* be embodied in written statements addressed to the person responsible for operating the installation, and *should* be filed with the permanent radiation safety record and, where appropriate, in the minutes of the Radiation Safety Committee proceedings.

5. Instrumentation

Radiation monitoring instruments may be divided into two classes: those for measuring radiation fields and those for measuring radioactivity. Several types of radiation detectors are to be found in instruments of both classes, but since the end use demands different features, these types will be discussed under each class. Information on radiation protection instrumentation is also presented in ICRU Report 20 (ICRU, 1971) and by Kiefer *et al.* (1969).

5.1 Instruments Used for Measurements in Radiation Fields

5.1.1 *General Properties*

5.1.1.1 *Properties of all Instruments.* Section 3.2.4 describes the factors to be considered when measuring radiation, including the type, energy spectrum, direction, intensity, and variation in space and time of the radiation. That section also describes the usual method of making such measurements: the instrument is calibrated in a known radiation field and then the field of interest is measured with the assumption that the calibration is valid. Most instruments have properties that may require modifications of this simple procedure.

Energy Response. If the energy spectrum of the radiation field of interest differs significantly from that of the calibration field, a correction may be necessary. When the same dose rate (or *exposure* rate, flux density, etc.) is produced by radiations of different energy spectra, instrument readings may be different because of differences in absorption, scattering, or rate of production of secondary radiations in the components of the instrument, or because of differences in the response of the radiation-sensitive element in the instrument. This property of an instrument is best characterized by measuring its response to several energies of monoenergetic radiation (Kathren *et al.*, 1971; Storm *et al.*, 1974). An example is shown in Figure 1, which gives such data for three gamma-ray measuring instruments calibrated with radium gamma rays (Curve A, Larsen *et al.*, 1954; Curves B and C,

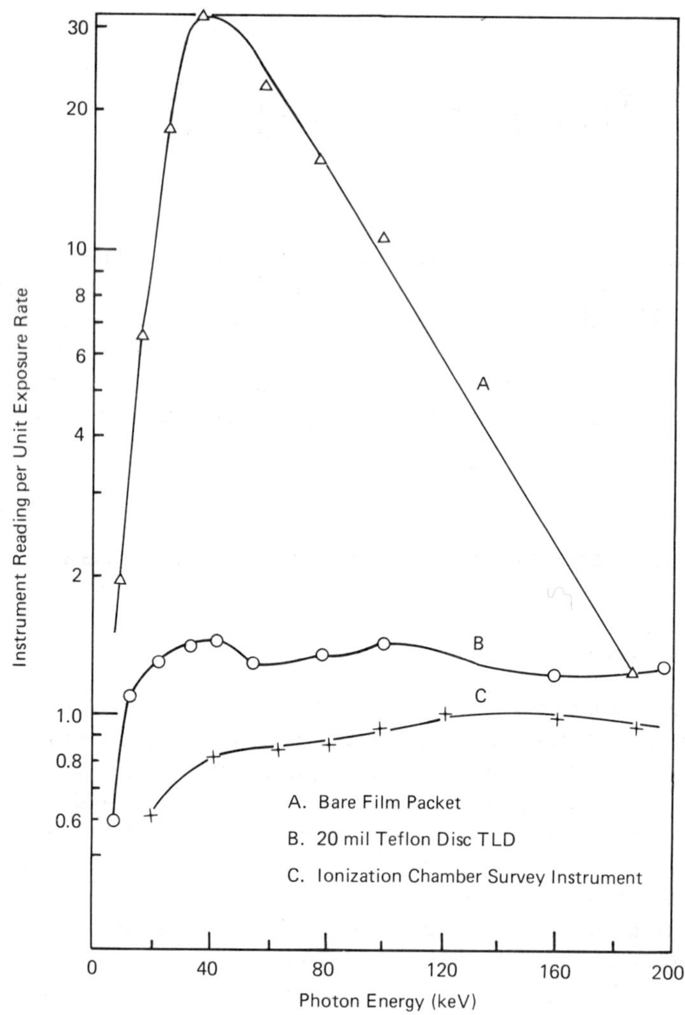

Fig. 1. Energy dependence of response of three typical x- and gamma-ray instruments.

Krohn *et al.*, 1969). It gives the quotient of the reading of the instrument for photons of different energies and the *exposure* rate (as measured absolutely with a free air chamber or a calibrated cavity chamber). If these curves were straight horizontal lines, the instrument would respond equally to all energies and could be used without correction in radiation fields having spectra different from that of the calibration field. It is frequently possible to obtain instruments that respond essentially equally to all energies for which they are to be used.

If a suitable instrument is not available, the course of action will depend on the magnitude of the dose equivalents involved. If the dose equivalents are small compared to the maximum permissible value, then measurement errors by a factor of 2 are acceptable. Hence, use of an instrument with poor energy response or a poorly-known correction for that response is acceptable if it leads to errors no greater than this. For dose equivalents close to the maximum permissible value, an accuracy of about 30 percent is desirable (Sections 4.2.1.3 and 4.3.1.4). If that accuracy cannot be achieved with the instrument, usually the simplest course is to obtain an instrument providing such accuracy. The alternative is to make measurements of the spectrum of the radiation of sufficient accuracy to permit correcting the reading of the instrument through use of its energy response curve (a curve such as shown in Figure 1). A crude description of the radiation spectrum may be adequate for this purpose. Accurate radiation spectrometry is, in general, a difficult undertaking, and the necessary instruments or techniques may be in a less developed state than those for the determination of the required quantity.

Directional Response. If the directions from which the radiations arrive at the instrument differ significantly from those in the calibration field, correction may be necessary. Ordinarily, all the calibration radiations come from very nearly the same direction. Other directions of incidence may produce different instrument readings because of differences in absorption or scattering in the components of the instrument (particularly in handles, circuit boxes, etc.). This property of an instrument is best characterized by measuring its response to the same radiation arriving from different directions. An example is shown in Figure 2, giving such data for a gamma-ray, a beta-ray, and a neutron instrument (Curve A, Kathren, 1967; Curve B, Roesch and Donaldson, 1958; Curve C, De Pangher, 1959). Plotted here is the quotient of the reading of the instrument for a source of radiation placed in the direction from which the calibration radiation comes and the reading as the instrument is turned through different angles. In the polar coordinate system used for this figure, equal response of the instrument to radiation from different directions is represented by a circle centered at the origin. The instruments shown do not respond equally in all directions, however, and most instruments do not. Indeed, it appears at first glance that the beta-ray instrument responds very poorly; however, this is a special case that will be discussed further in Section 5.1.2.2 below. With this exception, the instrument should respond essentially equally to radiation from all directions for which it is to be

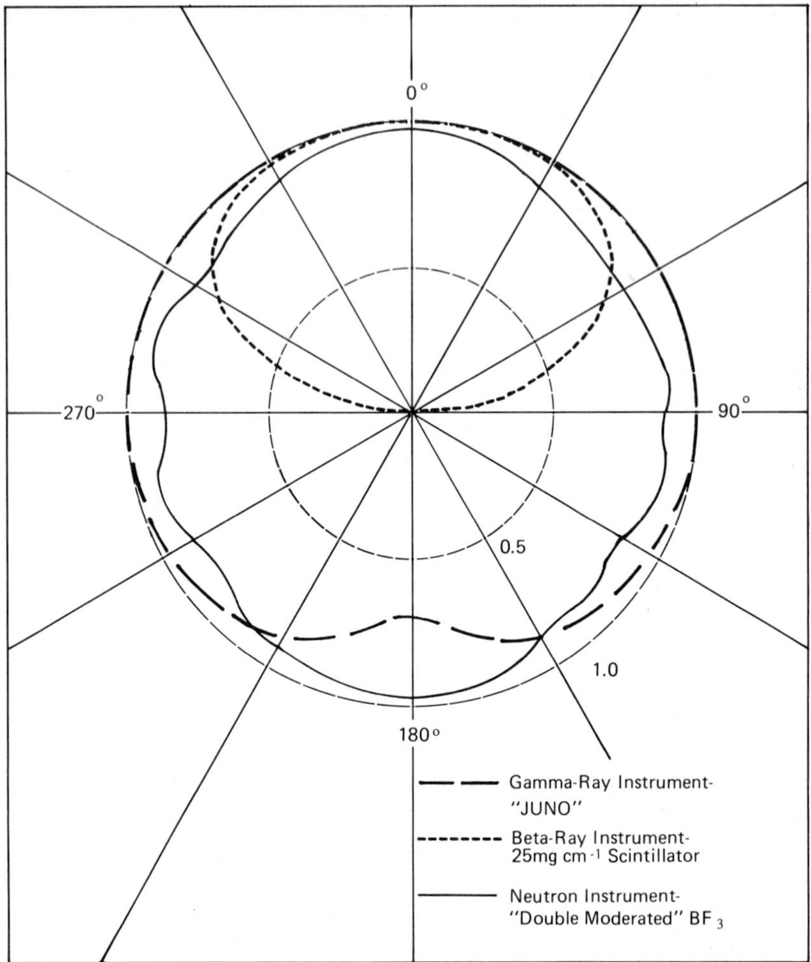

Fig. 2. Directional dependence of response of gamma-ray, beta-ray, and neutron instruments.

used. If only beams of radiation or small sources are in use, equal response within only a limited range of angles is acceptable. The course of action to be followed in other situations depends on the magnitudes of the dose equivalents involved, much as in making correction for poor energy response. If the dose equivalents are small, large errors are acceptable and it may not be necessary to make any correction; if the dose equivalents are large, more accuracy is required and it may be necessary to determine the directions of the radiations. Tests with the instrument itself to see if corrections for direction dependence of the response are necessary, may be misleading. If readings for different orientations of the instrument at a given point are significantly different, correction is needed. If they do not differ, however, correction may still be required. For example, if the instrument were in a perfectly isotropic field, it would give the same reading no matter how it was oriented and regardless of how poorly it might respond in different directions. In this context, a decision that a correction is needed *should* be based on knowledge of curves such as those in Figure 2 and of the distribution of directions of the radiation being measured.

Rate Response. In discussing rate response, it is necessary to distinguish between two classes of instruments: (a) those that measure dose, *exposure*, fluence, etc., i.e., quantities that are the total amount in a given time: these are called *integrating instruments* (or meters); (b) those that measure dose rate, *exposure* rate, flux density, etc., i.e., quantities that are rates: these are called *rate instruments* (or meters).

If the rate (dose rate, *exposure* rate, flux density, etc.) measured differs significantly from that in the calibration field, correction may be necessary. The rate may affect the instrument reading due to effects in the sensitive element of the device or due to limitations in the apparatus used to measure the output of the sensitive element. This property of an instrument is best studied by measuring the response to different rates of the same radiation. For integrating instruments, such a test will ordinarily show a range of rates for which the reading is independent of the rate. For rate instruments, such a test will ordinarily show a range of rates in which the reading is proportional to the rate. Ordinarily, an instrument *should* be used only within these rate ranges. If an instrument must be used at rates for which its response is substantially in error (usually very high rates), then the rate *should* be determined and correction made for the difference between the response at the particular rate and at the rate during calibration. It is sometimes possible to test for the necessity of a rate correction: in accelerator installations, the intensity of the primary beam can be varied and the response of the survey instrument studied. Inverse square attenuation, in the case of point sources, can also be used to provide a controlled change of rate and establish the need for a rate correction in the instrument. A few instruments will "jam" at very high rates; that is, they will cease to function and the reading will drop to zero or close to zero. If one knows that the reading should not be small, a reading close to zero is a warning that jamming is taking place. The danger of mistaking jamming for a true low reading is obvious.

It is particularly necessary to know the rate response of instruments to be used near machines that produce radiation in short pulses. The instantaneous rates during the radiation pulses are usually very high. Instruments for this work should respond essentially equally (in a test such as the one described in the preceding paragraph) to rates ranging up to the highest instantaneous rates to be encountered. Rate instruments used near repetitively pulsed machines ordinarily indicate the average rate, i.e., the rate averaged over the total time, both during and between pulses, since only this average is needed for radiation protection purposes. This average rate will be much lower than the maximum instantaneous rate, but it is the latter that sets the requirements on the instrument.

In general, not only one but all of the parameters discussed above may be different in use and during calibration, and the corrections may be interdependent. For example, the correction for different radiation energies might depend on the direction from which the radiation came. This would make the correcting process very difficult. An instrument whose response is essentially equal for all energies, directions, and rates in tests such as those described above, for the ranges for which it is to be used, is expected to have essentially equal responses for any combination of such conditions and therefore requires no corrections. This is a powerful argument for using such instruments rather than instruments that need correction.

Mixed Field Response. Since some radiations (such as neutrons) have a higher quality factor than others, separate measurement of the component radiations is recommended when a mixed field of radiations is present. This is done either by using instruments sensitive to one radiation and not to the other or by using two instruments that are sensitive to both but to a different extent. In either case, recommendations concerning the properties discussed above apply to these instruments also, with the additional recommendation that they apply for all radiations simultaneously. For example, a neutron instrument containing a proportional counter will have a certain response to neutrons of a given energy; it *should* retain this response in all gamma-ray fields in which the device is intended to be used. (Note, however, that in gamma-ray fields of very high dose rates, the counter will "jam" and the device will stop counting neutrons.)

Unwanted Response. Interference by radiations that an instrument is not supposed to measure constitutes one kind of unwanted response. Extracameral sensitivity (response to radiation by some part of the instrument other than the intended detector) is another. Geotropism and response to heat, light, rf radiations, mechanical shock, etc., are examples of unwanted responses not due to ionizing radiation.

Fail-Safe Provision. In order to prevent failure to detect the presence of a radiation field and thus create a potential for exposing people unknowingly, it is desirable that malfunctions of an instrument be readily recognizable or that they always result in readings that are too high. An instrument with this property is said to be fail-safe. For example, if the jamming discussed above is signaled by an indicator on the instrument panel, the instrument is fail-safe.

Precision and Accuracy. The precision (or reproducibility) of an instrument is characterized by the range of readings attained in repeated exposures under identical conditions. Precision is usually expressed as the standard deviation of a series of readings. Typically, standard deviations of a few percent are obtained. The accuracy of an instrument expresses the deviation of its reading from the true value of the quantity it is intended to measure. The accuracy is affected by the precision, the dependence of readings on energy, direction, rate, etc., and by the accuracy with which the calibration was performed. Improper calibration may introduce a bias. The accuracy of calibration *should* be within 5 percent. Desirable measurement accuracy is discussed in Sections 4.2.1.3 and 4.3.1.4.

Calibration. Instruments used for radiation protection are not absolute instruments; that is, they require calibration in a known radiation field or comparison with instruments whose response is known. Many users of radiation protection instruments must rely on the manufacturer to calibrate their instruments properly. Users *should* arrange a reproducible radiation field in which the instruments are placed and read frequently, at least quarterly. The reading error due to lack of precision is minimized by computing the mean of several readings. If changes in the mean reading are detected, the instruments *should* be recalibrated promptly.

5.1.1.2 *Properties of Survey Instruments.* Typically, survey instruments are rate instruments. However, an integrating instrument can be used to measure an average rate by leaving it at a given position to measure the total radiation in a known time. This is a useful technique for very low rates or for the average rate with short recurring pulses. For example, average rates can be measured by this method near machines that can be operated for short periods only.

Time Constant. An important characteristic of a rate instrument is the time constant. This is a measure of the time necessary for the instrument to attain a constant reading when suddenly placed in a constant radiation field. That time constants are not always negligible is due usually to slowness of the readout apparatus. Time constants are generally given as the time required to arrive at $1 - e^{-1}$ (i.e., 0.63) of the final reading. The time constant of a rate meter *should* be small, in order to limit the exposure of the person taking the readings, and in some situations to reduce the necessary exposure time of the radiation source (e.g., of radiographic equipment). Typical time constants of good rate meters are one second or less.

Portability. To facilitate rapid surveys of large areas, survey instruments should be easy to move and easy to read. Size, weight, power requirements, and method of reading affect the ease of use. Battery-powered devices that are small enough and light enough to be hand-held for long periods, and that present the reading on an attached indicator, are preferred. Some instruments (e.g., some neutron detectors) or the apparatus used in reading (e.g., scalers or multichannel analyzers) are so heavy that they can be carried only with difficulty, or on wheeled carts. This limits the location of their use. Power cables supplying apparatus or signal cables returning to read-out equipment similarly limit the usefulness of a survey instrument.

Detector Size. The choice of the volume of a radiation detector may be influenced by the intensity of the fields it is to measure, the nature of its response to radiation, and the limitations of electronic circuits used to measure this response. Portable ionization chambers, for example, often have a volume of about one liter to measure 1 mR/h with currently available electronics. Similar size limitations also arise with other types of detectors. In Section 3.2.4, the difficulty was mentioned of measuring radiation fields that change significantly over a distance equal to a dimension of the instrument. This problem might occur near a small radiation source or in a small beam emerging through a shield. In the latter situation, the field may be uniform over its cross section, but the cross section is not large enough to cover the detector. In these circumstances, an estimate of the radiation intensity in the beam can be made in the following way: Move the instrument around to locate the maximum reading in the field. Multiply this maximum by the ratio of the area of the detector perpendicular to the beam and the cross sectional area of the beam. This product will always be larger than the true value of the quantity being measured. A possible alternative method is to measure the *exposure* rate or dose rate at a greater distance from the source of the radiation where the field may be wide enough to encompass the detection volume. A better solution in non-uniform fields of large dimensions is to use a detector that is small enough to make field variations over its own dimensions negligible.

5.1.1.3 *Properties of Personnel Monitoring Instruments.* The measurements required from personnel monitoring of external exposure are of integrated quantities in a given time rather than of rates. A rate instrument can, however, be used to measure total quantities if the integration is done as described in Section 4.2.2.6.

Background, Leakage, and Fading. An important characteristic of integrating instruments is that they are able to register from the time they are activated until the time they are read. This has at least two consequences, both of which tend to change the reading of the instrument. First, the instrument is exposed at least to background radiation, in addition to the radiation of interest. A correction can be made for readings due to other causes by subtracting the reading of a duplicate instrument that has been exposed under the same conditions as the measuring instrument but not to the radiation of interest. The other consequence is a possible change in the reading with time due to some intrinsic property of the detector; illustrated, in the case of an integrating ionization chamber, by loss of charge through leakage over insulators. Almost every integrating instrument has some corresponding phenomenon that results in a change of the output reading from the value registered on the prompt readout. These leakage or fading effects *should* be small in the time that will normally elapse before reading; otherwise, a suitable correction *should* be made. These effects can be kept within a few percent over the integrating period except in very warm or humid climates (Becker, 1973).

Size and Readout Systems. Since personnel monitoring instruments are normally fastened to the clothing of the worker, they *should* be small and light in weight. Preferably, the weight *should* be only a few ounces. Generally, the size problem discussed for survey instruments seldom arises except when making beta-ray measurements. A considerable reduction in size and weight is ordinarily made by reading the detector in a separate device. The bulk of personnel monitoring is done with instruments of this type. However, with such devices the result is not continuously available during the progress of the work to guide the worker or the safety officer. The advantage of having such information during the course of work in which the permissible dose may be attained rather quickly has led to the development of instruments that can be read at any time or that signal when a preset reading has been reached. Those that can be read at any time employ non-destructive readout. "Non-destructive" means that the reading of the instrument can be observed repetitively without changing it. Non-destructive readout is also advantageous when performed on a separate instrument; in the event the reading goes off-scale, it is possible to change to a readable scale.

5.1.2 *Specific Fields*

5.1.2.1 *X-Ray and Gamma-Ray Fields*

Quantity Measured. For radiation protection purposes, the quantity desired in the case of x and gamma rays is absorbed dose expressed in rad. Because of the long history of use of air-equivalent ionization chambers in the measurement of x rays in the medical field, and the practical difficulties involved in fabricating ion chambers with tissue equivalent walls and gas, x rays and gamma rays with energies below a few MeV are commonly measured in terms of *exposure* or *exposure* rate.

In radiation surveys the *exposure* rate is measured at the position a worker may occupy. No phantom body is used to simulate the absorption and scattering of the body. In personnel monitoring the *exposure* is measured with small instruments at the surface of the worker's body. Absorption and scattering affect these measurements and contribute to the differences from estimates obtained during surveys. As discussed in Section 3.2.4, in both situations the *exposure* (in roentgens) is taken to be equal to the dose equivalent (in rem) for radiation protection purposes. Where the energy of the photons to be measured is greater than a few MeV, it is not possible to meet the electron equilibrium requirement for *exposure* measurements and therefore dose is usually measured.

Exposure and *exposure* rates are quantities characteristic of the effect of photons on air. Therefore, the materials of which the detector is made must be fairly close to air in atomic composition or else provide some way of altering the response of the detector so that it simulates that of air is necessary to make the response of the instrument the same to equal *exposures* from photons of all energies. This can usually be done satisfactorily if the energies of the photons to be measured are all in the energy range where photoelectric absorption predominates or all in the range where Compton scattering predominates. The dividing point between the two is roughly 0.05 MeV. This is often within the range of interest and necessitates special measures for correcting the instrument response.

Survey Instruments.

Ionization Chambers. Most of the x-ray and gamma-ray *exposure* rate measurements for the surveys discussed in Section 3 are made with small, portable ionization chambers. Usually electrometers are part of the portable device and show the *exposure* rate on a meter. Time constants are determined by the electrometer and are usually about one second. The total weight of these devices is usually less than five pounds. Ion chambers with separate readers are also available and are useful for measuring either very high or very low *exposure* rates that cannot be measured conveniently with ordinary survey instruments. The ion chambers are usually about one liter in volume when measurements in the range of a few mR/h are desired; this occasionally leads to the size problem discussed in Section 5.1.1.2. Ion chambers may also be placed at fixed locations where frequent or continuous *exposure* or *exposure* rate measurements are desired.

Ion chambers made of plastic or other organic or low atomic number materials usually give readings per unit *exposure* that are independent of photon energy except at energies below about 0.05 MeV (see Figure 1) (Krohn *et al.*, 1969). Below this energy the reading per roentgen may be several times larger or smaller than it is at higher energies, but the use of compensating materials in the chamber can reduce this effect considerably. Chambers designed for higher photon energies usually have walls so thick that they attenuate photons of less than about 0.03 MeV to the extent that a correction is required.

In order to measure *exposure* or *exposure* rate the air in an ion chamber must be surrounded by an "air-equivalent" wall material whose thickness exceeds the range of the most energetic secondary electron the photons can produce. Walls strong enough to be self-supporting are usually a few millimeters thick. This is thick enough to provide electronic equilibrium up to several MeV. This thickness is much more than needed for low energy x rays; thinner, supported walls are sometimes used to reduce absorption of x rays of a few tens of keV. The required wall thickness *should* be determined for the particular field. Figure 2 contains a typical curve showing the effect of direction of the radiation on an ionization chamber reading. As discussed in Section 5.1.1.1, no correction is required for most purposes. Ionization chambers are available for determining *exposure* rates of up to a few thousand R/h.

G-M Counters. G-M counters are used in surveys for the detection of x- and gamma-ray fields. Their dead time sets a limit to their counting rate. This generally limits their use to *exposure* rates in the range from background up to a few mR/h. Some halogen-filled G-M tubes (Fowler *et al.*, 1959) can be used as ion chambers when the counting action is no longer possible; these are useful for *exposure* rates up to about 1000 R/h. G-M counter instruments are available in portable units weighing only a few pounds, with the counting rate indicated on a meter or by audible clicks or rate-dependent tones. The recognition of a change in counting rate is determined by the ability of the user to recognize a change in click rate. The meter and variable-tone devices have response times of about one second.

The counters respond to the number of ionizing events within them and give no information about the energy associated with the events. Therefore, they do not respond with equal count rates to equal *exposure* rates from photons of different energies. It is for this reason that they are generally used only for detection rather than measurement. Their response can be made reasonably independent of photon energy in the range 200 keV to 3 MeV (Krohn *et al.*, 1969; Sinclair, 1950).

G-M counters are sometimes used to estimate *exposure* rates where the rate is low enough for the resulting inaccuracy of its measurement to be unimportant, or when the photon energy is the same as that used for calibration. For example, they may be used to measure low-level leakage radiation from devices containing sealed gamma-ray sources for radiography or teletherapy.

Scintillation Instruments. Scintillation instruments may also be used for x-ray and gamma-ray surveys. In relatively weak radiation fields, while the associated electronic parts of the instruments cause their overall size to be about the same as that of ion chambers, the detecting volume can be much smaller. A one-cubic-centimeter crystal is often adequate; the higher sensitivity of larger crystals permits the survey of areas at near-background *exposure* rates.

Scintillation devices have count-rate limitations due to the duration of the light flashes, but they can count much faster than G-M counters. The counting rates are higher than those in G-M counters for the same field, and therefore these instruments are useful for locating and delineating weak x-ray and gamma-ray fields.

Organic scintillators are close enough to air in effective atomic number to require little correction for energy dependence except at energies below about 0.1 MeV. In anthracene, for example, the response per roentgen falls primarily because of lower energy absorption in these materials relative to air (Ramm, 1966). Incorporation of a small amount of high-Z material in the crystal can partially offset this drop (Belcher and Geilinger, 1957).

NaI(Tl) crystals, widely used in gamma-ray spectroscopy, make very sensitive detectors; however, the count-rate per unit *exposure* falls rapidly as photon energy increases. For this reason, these counters cannot be used for making *exposure* measurements without specific energy calibration or applying energy correction factors. They are best used for radiation detection only, with subsequent measurements carried out with other instruments.

Personnel Monitoring Instruments.

Ionization Chambers. Small ion chambers, about the size of pencils, are often used for personnel monitoring. Most of them are charged and read on a separate device, but chambers with built-in electrometers are also available. These small chambers are ordinarily used to measure a few tens to a few hundred milliroentgens with electronic or quartz-fiber electrometers. Special electrometers can increase the sensitivity to permit measuring a few milliroentgens. Adding capacitance to the chambers reduces the sensitivity and permits measurement of higher *exposures*. Still smaller, less sensitive chambers are worn on the wrist, finger, etc. The remarks in Section 5.1.2.1 about wall material and wall thickness of ionization chamber survey meters apply here also.[16]

Photographic Film. Small badges containing special x-ray films are popular personnel dosimeters. The sensitivity of available emulsions is sufficient to detect about 10 mR of cobalt-60 gamma radiation and a few mR of diagnostic x rays, and is therefore high enough for radiation protection purposes. The useful range is great enough to measure most emergency *exposures*. The range from a few mR to 2000 R can be covered by two commonly available films or two emulsions of different sensitivity on one film base.

The optical density per roentgen is essentially constant for x and gamma rays with energies above 200 keV, if electronic equilibrium is established at the film. For energies below 200 keV, however, the photographic density per roentgen increases and is 20 or more times higher for photons of about 40 keV than for 1 MeV photons. For dosimetry below 200 keV, metallic filters covering the film reduce the energy dependence. Filters are most widely used to provide an estimate of the photon energy from the ratios of the densities obtained behind filters and at an open window (Herz, 1969). Thin plastic filters permit differentiation between x-ray and beta-particle doses.

Photographic film has several undesirable characteristics. Fogging may result from mechanical pressure, elevated temperatures, or exposure to light before development. Fading of the latent image results in a reduction of density which is dependent on the time interval between exposure and development and the degree of exposure to moisture and heat. This may be minimized by special packing to exclude moisture from the film and by storage in a refrigerator or freezer prior to distribution (Cheka, 1954; Becker, 1973). Film dosimeters also exhibit directional dependence, particularly for the densities recorded behind metal filters.

Considerable care must be exercised in developing the films and in preparing calibrated film for comparison (Ehrlich, 1954; Becker, 1966a).

Thermoluminescence. Thermoluminescent dosimetry has consider-

[16] Performance specifications for pocket dosimeters are published by ANSI (1972a).

able potential for application in the field of personnel monitoring. It is rapidly displacing film dosimetry as the most common method (Attix, 1972). The desirable characteristics include its wide linear range, short readout time, relative insensitivity to field conditions of heat, light, and humidity, reusability of the dosimeters, and for some phosphors, energy independence. Response is also rate-independent up to 10^{11} R/s.

Practical personnel dosimeters are available that employ LiF:Mg, Ti (Cipperley, 1966; Gollnick, 1972), CaF_2:Mn (Brooke and Schayes, 1966), $CaSO_4$:Dy (Webb *et al.*, 1972), and $Li_2B_4O_7$:Mn (Thompson and Ziemer, 1972). LiF is the most commonly used phosphor, usually in the form of crystal chips or impregnated Teflon (Cox and Lucas, 1974). Other systems include a type using a sealed glass envelope with the phosphor bonded directly on a heater wire and a type using a hollow glass needle filled with the phosphor that can be read by externally heating the needle. The LiF and $Li_2B_4O_7$:Mn phosphors are nearly tissue-equivalent even at low photon energies. The response of LiF is linear up to about 1000 R, above which it becomes supralinear. Careful annealing before and after irradiation and reading is necessary. Characteristics such as the fading and shape of the glow curve may change with annealing temperatures and times. The CaF_2:Mn phosphor is sensitive down to a few mR and has a linear response up to 10^6R. It exhibits an increasing response as photon energies are reduced below 100 keV. Thermoluminescent dosimeters can be used with energy compensation shields to reduce their energy dependence. The precision of individually calibrated detectors is about $\pm$ 3 percent with better than $\pm$ 10 percent for a group of detectors from the same phosphor batch.

Very small thermoluminescent detectors (10 mg to 1 g) can be used conveniently to measure *exposure* to specific parts of the body. They probably represent the method of choice for measurement of finger, hand, or eye dose.

Radiophotoluminescence. Silver-activated metaphosphate glasses are available in which radiation produces fluorescent centers causing light of an orange color to be emitted when the glass is irradiated with ultraviolet light. The intensity of the fluorescence is proportional to radiation exposure up to 1000 R or more. Self-contained readers with ultraviolet light source and a photomultiplier tube to measure the fluorescence are available (e.g., Dade *et al.*, 1972), and are capable of measuring *exposures* down to 10 mR for the most sensitive glasses (Yokota and Nakajima, 1965).

Glass dosimeters exhibit energy dependence for photon energies less than 200 keV. The dependence is greater in the "high Z" glasses employing aluminum, barium, and potassium phosphates, and smaller in the "low-Z" glasses employing aluminum, boron, magnesium, sodium, and lithium phosphates. With proper filtration, this energy dependence can be corrected (Piesch, 1972).

The amount of luminescence produced by a given *exposure* depends on the time interval before reading; i.e., "fading" occurs (Becker, 1973). The readout is nondestructive. For maximum sensitivity, contamination of the glass by dust or grease must be eliminated by careful cleaning. The dosimeters may be annealed by heat treatment and reused.

Calibration

Laboratory-standard instruments for measuring *exposure* from x rays generated at exciting potentials in the range from 10 to 250 kVp are calibrated by the National Bureau of Standards (NBS) for a number of selected beam qualities specified in terms of filtration and half-value layer, by comparison with a standard free air chamber. Such calibrations are offered for limited instrument ranges and *exposure* rates. For use with photons of higher energies, laboratory-standard instruments for measuring *exposure* with full scale reading from 0.1 R to 100 R, or instruments designed to measure *exposure* rate in a range from 0.01 to 15 R/min, can be calibrated at NBS by comparison with either a cesium-137 or a cobalt-60 source that, in turn, was calibrated by absolute measurements with cavity-ionization chambers.

The National Bureau of Standards also calibrates sealed radioactive sources containing up to 250 mg of radium-226 in equilibrium with its decay products, and sealed cobalt-60 gamma-ray sources giving *exposure* rates between about 0.5 and 200 mR/h at a distance of one meter. These calibrations are given in terms of *exposure* rate.

Instruments calibrated at the National Bureau of Standards for particular radiation qualities may be used to calibrate other laboratory sources that provide radiation of similar quality. These sources, or the sources directly calibrated by NBS, may then be used to check the calibration of further laboratory instruments. Inverse-square law tests *should* be performed in every new set-up to check on exposure geometry. If the response of the instruments to be checked depends strongly on radiation quality, care *should* be taken to keep scattered radiation from reaching the instrument by preventing the beam from striking more extraneous material than is unavoidable. If the radiation source is of the collimated type, it is usually helpful to limit the beam diameter in the plane of the instrument to a size not much larger than twice the largest linear dimension of the instrument. The adequacy of a given source-to-instrument and instrument-to-scatterer distance varies with the geometry of the layout and the type of instrument. Examples of how the reading of a 25 R condenser-type ionization chamber exposed at a distance of 70 cm from a lightly filtered, well-collimated, 250 kVp x-ray source is influenced by various scattering materials in its vicinity are given in ICRU Report 10e (ICRU, 1963). For instance, when the chamber is exposed on a ½ inch (1.27 cm) thick sheet of Masonite which, in turn, lies on an aluminum table, it is shown to read 15 percent more than when exposed free in air. At a distance of 1 meter from a cobalt-60 gamma-ray source suspended in the center of a 15 foot × 15 foot × 9 foot (4.6 m × 4.6 m × 2.7 m) room with thick concrete walls, a similar chamber will read about 10 percent higher than if it had been exposed in a scatter-free geometry (Wyckoff, 1950). For a device with a more strongly energy-dependent response, such as photographic film, the change in reading would be larger, both for cobalt-60 gamma radiation and for low energy x radiation.

5.1.2.2 *Beta-Particle and Low-Energy Electron Fields.*[17]

Quantity Measured. Electrons of about 2 MeV or less in energy have such short ranges in solids that they are absorbed by the surface layers of the body. The outer part of the skin is an inert layer not affected by the radiation; therefore, measurements are required only at depths in the body equal to (or greater than) the thickness of this inert layer. By convention this thickness is taken to be 7 mg/cm^2. The quantity measured at this depth is the absorbed dose or the absorbed dose rate.

To measure the dose at this depth, part of the wall of the instrument, the entrance window, must be 7 mg/cm^2 thick. The entrance window is usually a thin, electrically-conducting plastic sheet. The detector behind the window is usually a broad, thin layer of material. The detector cannot be very thick, since absorption of the electrons within it causes the instrument to measure the average dose (rate) throughout the detector, whereas the surface dose is usually required. Thicknesses *should not* exceed 10 mg/cm^2 (about 10 cm of air). The extension of the detector in the direction perpendicular to the field is limited to prevent averaging the dose over too great a cross section of the field. The base on which the window and detector are mounted is usually sufficiently thick to simulate effectively a human body with respect to absorption and backscattering of the electrons. Therefore, these in-

[17] This section excludes discussion of tritium measurements, that is included in Sections 5.2.1.3, 5.2.2.1, and 5.2.3.2.

struments detect only the electrons that arrive from sources in front of the entrance window. Electrons arriving nearly tangentially to the entrance window must penetrate a long, slanted path through the window and many of them will be largely or completely absorbed in it. These are the reasons for the dependence of the instrument response on the direction of the beta-particle radiation, shown in Figure 2.

Survey Instruments

Ionization Chambers. Most survey measurements of beta-particle and low-energy electrons are made with small, portable ionization chambers similar to those used for x-ray and gamma-ray surveys. Chambers made especially for electron measurements have a very small ratio of chamber thickness to diameter of the entrance window, for the reasons given above. More often, however, measurements of electrons or beta-particles are made with ionization chamber instruments designed primarily for measuring x or gamma rays. One side of the chamber comprises a thin conducting plastic sheet that is covered, when measuring photons, with a thick piece of the chamber material; the thick cover is removed for measuring electrons. The ratio of depth to diameter of these chambers is usually large, with the result that some parts of the chambers cannot be reached by the electrons from a source of large area; therefore, the response depends strongly on the direction of the electrons. For small electron sources on the axis of the entrance window the error is minimal, but for extended sources, such as one might find in a grossly contaminated area, the error may be large. Another technique used for adapting x-ray or gamma-ray instruments to electron measurements is to remove almost the entire wall of the detector and replace it with a very thin wall. Such a chamber will measure a dose (rate) that is higher than a person would receive, because it does not geometrically simulate the body surface as does the beta-ray instrument with the response shown in Figure 2.

The walls of an ionization chamber for electrons *should* be made of materials similar to tissue in composition, but the exact composition is not critical. In the case of ionization chambers for x or gamma rays, the atomic composition of the walls is important because it plays an essential role in the conversion of photon energy into the energy of secondary electrons. However, for electrons and beta rays the function of the walls is merely to simulate the absorption and backscattering by the body. The remarks about size, sensitivity, response time, and readout methods for x-ray and gamma-ray ionization chambers also apply to those for electrons.

G-M Counters. G-M counter survey instruments for electrons are similar to those for x and gamma rays except that the walls must be sufficiently thin for the electrons to penetrate to the sensitive volume. If the counter wall is thin enough for this purpose and is also provided with a cover that is sufficiently thick to stop the electrons, the difference in readings with the cover on and off can be used to distinguish between electrons and x or gamma rays. Since the counter provides a record of the number of particles rather than of energy deposited, its reading is only approximately proportional to dose (since LET varies slowly with electron energy). Its chief use is to detect and establish the relative magnitude of electron fields.

Scintillation Instruments. Very thin layers of scintillators, particularly organic scintillators, behind a thin entrance window, and coupled to photomultipliers from which the current is measured, constitute good electron dose meters. Because the scintillator is thin, the total light output, and hence the current, is small. They can, however, measure dose rates above a few mrad/h. Thicker scintillators produce more light, but their readings bear less relation to the surface dose rate usually desired. Scintillation counters with thin entrance windows are usually as sensitive to electrons as G-M counters of the same size. They, too, can be useful in locating electron fields. However, a covering opaque to light is necessary to exclude visible light from the scintillator and this may be sufficiently thick to provide significant or complete absorption of low-energy electrons.

Personnel Monitoring Instruments

Ionization Chambers. Small ionization chambers, similar to those used for x and gamma rays, have been adapted to electron measurements by the use of thin entrance windows. Usually a group of perforations is made in the wall of a chamber and covered with thin plastic to make a number of small entrance windows without reducing the strength of the chamber wall. This results in the sensitivity of the chamber for electrons being less than that for x and gamma rays—roughly in the ratio of the window area to the total surface area of the chamber. The other characteristics of the chamber are similar to those of chambers used for x and gamma rays.

Photographic Film. In similar fashion, film badges meant for measurement of x and gamma rays have been applied to electron measurement by the use of a thin entrance window (often the "open window") over part of the badge. It is not generally possible to make such a window only 7 mg/cm^2 thick; thicknesses of about 100 mg/cm^2

are typical. If there are two pieces of film or two emulsions on a single film in the badge, the average thickness of material over both emulsions is even greater. For a given electron or beta-particle energy, the absorption of electrons in this additional thickness of material can be corrected for in the calibration process. However, if measurements are made at other energies, special corrections are required.

In other respects, the response of film badges to electrons and beta particles is similar to that for high energy x and gamma rays. In particular, electrons produce about the same optical density as an equal dose of x or gamma rays. If, however, the additional absorption correction is included as described above, the sensitivity for corrected surface dose is considerably reduced. For example, if the film is exposed under 100 mg/cm^2 of material and its reading ascribed to the radiation present at 7 mg/cm^2, the optical density per unit dose is one-half or less than if the optical density were ascribed to the actual dose in the emulsion(s).

Thermoluminescence. Discs of thermoluminescent material a fraction of a millimeter thick, e.g., Teflon impregnated with LiF, may also be used to measure low energy electrons. Because of the rapid absorption of the electrons in the discs, the response per rad is less than that obtained with high-energy gamma rays. Correction for the non-uniform dose distribution can be made; otherwise, as for film, calibration for specific electron energies *should* be performed.

Calibration. The basic instrument for measuring absorbed dose in radiation fields exhibiting strong spatial variations, i.e., in fields of radiation that are heavily attenuated, is an extrapolation chamber. The basic method for calibration of electron and beta-particle instruments is comparison with a suitable extrapolation chamber in an electron field. Extrapolation chambers are, however, delicate laboratory instruments. Only a few large laboratories can be expected to maintain them. Several measurements of the dose rate in tissue have been made with extrapolation chambers in contact with thick slabs of natural uranium in equilibrium with its daughter products. This electron field can be considered as a standard. The dose rate is 225 mrad/h at a depth of 7 mg/cm^2 in tissue equivalent material.

Another technique is to immerse the instrument (covered with a very thin, water-tight wrapper) in a solution containing a beta-particle emitter. At points in the liquid at distances that are greater than the ranges of the beta particles from any boundary of the liquid the dose rate is 2.13 CE rad/h, where C is the concentration of the emitter in the solution in units of μCi/g and E is the *average* energy of the beta particles in MeV (about one-third their maximum energy). If the instrument simulates the body (i.e., the back of the instrument is thick enough to prevent the penetration of electrons to the sensitive volume) the dose rate at the sensitive volume is one-half 2.13 CE times a factor giving the attenuation in the entrance window. The latter can be estimated by measuring the absorption in added layers of the entrance window material.

Usually instruments for electron measurement are calibrated with x or gamma rays. The calibration for *exposure* rate in R/h is so nearly equal to the desired calibration for dose rate from electrons in rad/h that the difference is generally ignored in the calibration of instruments for radiation protection purposes.

5.1.2.3 *High-Energy Electron Fields.* The same instruments used for x-ray and gamma-ray measurements are used for measurement of high-energy electrons. Frequently it is necessary to cover the survey instruments with additional tissue-like material to place their sensitive volumes at the depth of maximum dose. Electrons with energies of more than about 2 MeV have ranges sufficiently long to affect tissues below the surface layers of the body. These high-energy electrons usually produce their maximum absorbed dose at some depth below the surface of the body. At 2 MeV this maximum is about twice the dose at the surface, while above 5 MeV it decreases to a value 10 to 20 percent above that at the surface. For radiation protection purposes, the calibration of these instruments in rad in tissue is assumed to be equal to the calibration for exposure in roentgens, as noted in Section 5.1.2.2. above.

5.1.2.4 *Neutron Fields.* There are many difficult problems associated with neutron dosimetry. Since many of these are dealt with in NCRP Report No. 25 (NCRP, 1961a) and ICRU Reports 10b and 26 (ICRU, 1964, 1977), the treatment here will be brief. Frequently the chief problem is to measure neutrons and gamma rays separately in the same radiation field. Neutrons are almost always accompanied by gamma rays. Neutron doses *should* be measured separately from gamma-ray doses, because the neutrons have a quality factor from two to ten times that for the gamma rays (NCRP, 1971a).

In most materials, neutrons are not strongly absorbed until they slow down to thermal energies. They can still produce nuclear reactions at these energies and release charged particles or gamma rays or produce radioactivity. In practice, it has been found that in areas potentially occupied the absorbed dose due to radioactivity is small compared with the doses from fast neutrons and gamma rays. Although instrumentation for thermal neutrons is simple, problems arise in the proper calibration.

Quantities Measured. The dose equivalent due to neutrons is usually deduced in one of the following ways: (1) The distribution of absorbed dose in LET is measured, each part of the distribution is multiplied by the appropriate quality factor (Q), and the products are summed to give the dose equivalent. This technique is lengthy and is usually used only where the neutrons produce a major share of the absorbed dose. However, since it makes use of the values of Q selected by the NCRP rather than high, limiting values (as in methods (3) and (4)), it gives the best determination of the dose equivalent. (2) The neutron dose equivalent is measured with an instrument that primarily responds to the kerma in tissue or with an instrument that primarily records thermal neutron fluence after empirical weighting by a surrounding moderator. (3) The absorbed doses due to the neutrons and to other radiations (particularly gamma rays) are measured separately by any of the techniques described below. Then the neutron absorbed dose is multiplied by 10 (the upper value of the applicable Q) to give its dose equivalent. (4) The total absorbed dose from all radiations is measured and multiplied by 10 to give the dose equivalent. This method usually gives a considerable over-estimate of the dose equivalent, since it applies the highest Q to *all* of the radiations.

In radiation surveys of neutron fields the kerma rate (or its distribution in LET) or the dose equivalent rate at the position to be occupied by the worker is the quantity usually measured. It is assumed that the absorbed dose rate in the worker will equal the kerma rate. For personnel monitoring, the absorbed dose at the surface of the body is measured.

Survey Instruments

Ionization Chambers. One method of determining the fast neutron kerma in a mixed neutron-gamma ray radiation field is to use two instruments with different neutron sensitivities and equal gamma ray sensitivities (NCRP, 1961a). The difference in the readings of the devices is due to the neutrons. The principal system using this concept employs ionization chambers in which the first chamber is made of a plastic whose composition is close to that of average tissue and is filled with a gas of the same composition as the plastic, while the second chamber is made of graphite or conducting Teflon, and is filled with carbon dioxide. The sensitivity of the latter for neutrons is 8 to 24 percent of that of the tissue-equivalent chamber. For radiation protection purposes it is adequate to make an average correction of 15 percent. Several other combinations of chambers have been used, e.g., identical metal chambers, one filled with methane and the other with carbon dioxide. A system in which the wall material and the gas are tissue equivalent has the advantage of requiring no correction for different energies, and its response is directly proportional to the kerma in tissue at the tissue surface.

Ionization chambers whose inner surfaces are lined with a boron compound are very sensitive to thermal neutrons because of the (n,α) reaction in the boron. Readings from a similar, unlined chamber are used to correct for the effects of gamma rays. Chambers of this type can be used in portable instruments for surveying, but usually are used as condenser chambers. The chamber is charged, exposed, and then read on an electrometer.

Proportional Counters. Proportional counters are employed in several ways to measure neutrons. A proportional counter with polyethylene walls and filled with ethylene or cyclopropane is frequently used in surveys of neutron fields (Hurst, 1954; Wagner and Hurst, 1958). The quantity actually measured is the kerma in ethylene, which is approximately proportional to that in tissue at all energies. The proportionality factor is 1.34 for ethylene and 1.45 for cyclopropane. The measurement is made by obtaining a pulse-height distribution of the counter pulses. The sum over all pulse-height intervals of the product of pulse height and number of counts is proportional to the kerma. The counts due to gamma rays all fall at low pulse heights and a correction for their contribution to the neutron dose can be made by extrapolating the counts due to neutrons to zero pulse height. Since this measuring device requires a pulse-height analyzer and other specialized electronics, it is not employed as a portable instrument. Instruments of this type have been designed to indicate dose-equivalent (see page 109).

A spherical tissue-equivalent proportional counter (Rossi and Rosenzweig, 1955) has been developed to measure the LET distribution of the kerma from which the dose equivalent can be derived. With a spherical counter for particles having a single value of LET the pulse height spectrum can be predicted theoretically. This enables one to deduce the LET distribution from complex pulse-height spectra. Multiplying the LET distribution by the appropriate Q and summing the products (integrating) leads to the dose equivalent.

Fast-neutron survey instruments have been developed that are based on slowing down (moderating) the fast neutrons and measuring the resulting low-energy neutrons. A BF_3 proportional counter surrounded by the proper thickness of paraffin or polyethylene gives a counting rate proportional to the neutron kerma rate (De Pangher, 1959; Ladu *et al.*, 1965; Rotondi, 1966). By proper selection of its sensitivity, such a counter can be used for rates as low as those due to cosmic ray background and up to those as high as encountered in reactor-shielding studies. The interference due to gamma rays is very small; gamma-ray *exposure* rates of several hundred R/h can be tolerated. Instruments of this type have also been developed to indicate dose equivalent (see later in this section).

Scintillation Instruments. Another detector capable of measuring fast neutrons for radiation protection purposes is comprised of spheres of polyethylene with a LiI(Eu) scintillation crystal mounted at the center of the sphere (Bramblett *et al.*, 1960; Lawson and Watt, 1964). Crystals highly enriched in ^{6}Li have a high detection efficiency because of the large (n,α) cross section, and provide effective discrimination against gamma rays because of the large Q value (4.8 MeV). A light pipe extends from the scintillator to a photomultiplier mounted outside the sphere. A determination of the approximate spectrum and dose can be made from measurements taken with different diameter spheres, if the approximate shape of the spectrum is already known. For a single sphere of 10-inch (25.4 cm) diameter, the counting rate is proportional to the neutron kerma rate and is essentially independent of energy. This type of detector has also been developed into a neutron dose-equivalent instrument (see later in this section).

Various other neutron detecting instruments employing scintillation counters have been developed and are available in portable form. An example is the ZnS-lucite detector (Hornyak, 1952). Care must be taken in interpreting its readings because the response of this instrument does not parallel that of tissue as a function of neutron energy, and therefore the instrument does not measure the kerma unless it has been calibrated with neutrons of the same energy as those being measured. Since high sensitivity can be achieved with this type of counter, it is useful for delineating neutron fields.

Activation. Neutron fluences may be determined by exposing samples of various materials and measuring the resultant induced activity. The sample may be activated by (n,γ), (n,f), (n,p), (n,α), or other nuclear reactions that may or may not have a neutron energy threshold. The activity is measured either by absolute counting methods or by comparative counting against a physically similar sample containing a known amount of the same radionuclides. From the published cross-section data and other parameters including the sample mass, neutron energy, and the counter characteristics, the neutron fluence can be calculated from the measured activity.

Activation systems are applicable in fields that are too high for measurement by count-rate systems and operate in the integral mode. The use of threshold detectors will permit a rough determination of the neutron spectrum. For fission neutrons, subsequent calculation of absorbed dose may be made with an accuracy of about 25 percent. For these reasons activation detectors of the threshold and non-threshold type are widely used in nuclear accident dosimetry and for surveys of high-level neutron fields.

For the detection of slow neutrons various metallic foils are used. Indium and gold, which exhibit high activation cross-sections for thermal and slow neutrons, are the most commonly used foils. In order to determine the thermal neutron fluence, a pair of foils is usually exposed, one of which is covered with cadmium. The difference in the counting rates measures the thermal neutron fluence, and the ratio of the counting rates, called the cadmium ratio, is a measure of the degree of thermalization in the neutron spectrum. Indium has large resonances in the intermediate energy region that may lead to erroneous estimates of thermal neutron fluence. Such foils, when surrounded by a hydrogenous moderator, can also be employed for fast neutron measurement.

Cobalt discs are useful detectors for determining exposure to neutrons over extended periods of time (Smith, 1961). If the natural cobalt disc is placed in a polyethylene moderator, the dose from a year's exposure can be calculated by counting the cobalt-60 in the disc. Because of its high sensitivity and the long half-life of cobalt-60, this system is well suited for radiation exposure evaluation.

Various threshold reactions are used in the dosimetry of fast neutrons. If the resultant active material is a gamma emitter, pulse-height analysis is usually employed for the measurement of specific gamma rays. Some of the more commonly used reactions are listed in Table 1.

These reactions generally have too low a cross section to be useful for radiation protection purposes.

One typical threshold detector system useful for doses in excess of about 20 rad contains foils of ^{239}Pu, ^{237}Np, and ^{238}U shielded with 1 to 2 g/cm^2 of ^{10}B and a ^{32}S detector, that provide thresholds of increasing level, while gold foils, with and without cadmium are provided to determine thermal neutron fluence (Reinhardt and Davis, 1958). It is

TABLE 1—*Some threshold reactions used in fast neutron dosimetry*

Reactions	Approximate Threshold (MeV)	Type of Counting	Half-life of Product
^{239}Pu(n,f) F.P.	0.01	Gamma	[a]
^{237}Np(n,f) F.P.	0.6	Gamma	[a]
^{238}U(n,f) F.P.	1.5	Gamma	[a]
^{32}S(n,p)^{32}P	3	Beta	14.2 d
^{46}Ti(n,p)^{46}Sc	5.5	Gamma	84 d
^{27}Al(n,α)^{24}Na	7.5	Gamma or beta	14.9 h
^{127}I(n,2n)^{126}I	11.0	Gamma	13.1 d
^{90}Zr(n,2n)^{89}Zr	14	Coincident Gamma	78 h

[a] Half-life depends on the time between irradiation and counting.

necessary to analyze the fission detectors within an hour or two after exposure because of rapid decay.

Dose Equivalent Instruments. Several varieties of instruments that give readings proportional to dose equivalent over a wide range of neutron energy are now commercially available. That is, the instrument reading intrinsically compensates for the variation in Q with neutron energy (ICRU, 1971, Appendix B). Such instruments are often called "rem meters" or "rem counters". They are of two types: (a) the "recoil" type, employing a proportional counter filled with a hydrogenous gas and responding to proton recoils from a hydrogenous liner in the counter; and (b) the "moderator" type, employing a thermal neutron detector surrounded by a cylindrical or spherical hydrogenous moderator.

The first type has the advantage of being light in weight and providing good gamma-ray discrimination, but its response falls off for neutron energies above 10 MeV and is zero for neutrons below the typical bias energy of about 200 keV. Readings proportional to dose equivalent are maintained for neutrons between these limits (Dennis and Loosemore, 1961; Andersson, 1961; Murthy *et al.*, 1967).

The second type takes many forms since any type of thermal neutron detector may be used, including BF_3 and ^{3}He proportional counters, LiI(Eu) scintillators, thermoluminescent materials, and activation foils. Moreover, the moderator may be in one or several layers, may be of differing sizes and shapes, and may contain thermal neutron absorbers (Nachtigall and Burger, 1972). In its original form (Bramblett *et al.*, 1960) a LiI(Eu) detector was located at the center of a single sphere of polyethylene. With this detector there are serious gamma-ray discrimination problems (Piesch, 1969) but these are reduced by employing a crystal enriched in ^{6}Li or preferably a proportional counter (Hankins, 1968; Rotondi and Geiger, 1968). The over-response of simple moderator types for intermediate energy neutrons may be reduced by including layers of thermal neutron absorber (e.g., boron-loaded plastic or cadmium foils) at selected positions within the moderator (Andersson and Braun, 1964; Leake, 1968). Recent developments have included the use of several moderators or several detectors (Nachtigall and Burger, 1972).

These instruments are very useful for neutron surveys and provide dose equivalent readings reasonably independent of neutron energy in the range from 0.2 to 7 MeV (e.g. ± 15 percent) with moderators of about 10 inches (25.4 cm) in diameter. Larger moderators improve the high energy performance. Response is typically too high by factors up to 2 for neutrons of the order 10 keV energy. When instruments in this class are used, it is important to know approximately the spectrum of neutron energies involved and the energy dependence of the instrument. Other disadvantages of these instruments are their weight, typically 10 or more kilograms, and the possibility of a misleading response in some designs in gamma-ray fields of high intensity.

Personnel Monitoring Instruments

Ionization Chambers. For neutron dose measurement, chambers in which both the gas and the wall are tissue-equivalent may be used. For thermal neutron monitoring, chambers the size of pencils and lined with a boron compound, typically with a thickness of 1 mg/cm^2 of boron, are available. In these chambers, and in similar chambers lined with uranium, ionization is produced by the secondary particles produced in (n,α) and (n,f) reactions, respectively. The slow neutron sensitivity is high but the chambers are also sensitive to gamma rays.

Photographic Film. The effect upon normal photographic emulsions per unit of absorbed dose produced by heavy charged particles (such as recoil protons set in motion by fast neutrons) is much smaller than that produced by x or gamma rays. The sensitivity of x- and γ-ray monitoring film to fast neutrons, for example, is only 1 to 10 percent of its gamma-ray sensitivity. Because of the high LET, much of the energy absorbed is wasted on silver grains that are already developable. To overcome this effect, special nuclear emulsions with relatively small grains (nuclear track film) have been produced that exhibit developable grains when exposed to heavy particles but are relatively insensitive to x or gamma rays. The path of the particle as it passes through the emulsion can be observed microscopically as a line of developed grains (or a track). Dosimetry of fast neutrons is accomplished by counting the tracks produced by proton recoils from the hydrogenous gelatin matrix and film base (Cheka, 1954). Nuclear track film is used extensively as a personnel dosimeter and also for area surveys around reactors and accelerators.

Fast neutron dosimetry is facilitated by surrounding the nuclear emulsion with tissue-equivalent material of such thickness that the number of tracks produced in the emulsion per unit neutron dose is essentially independent of the neutron energy. In this case, the number of tracks in a given volume is taken to be proportional to the absorbed dose. In practice, the number of tracks in a given area (typically 25 to 50 microscopic fields of view) is measured. The fast neutron sensitivity is about 10^4 tracks/cm^2-rem depending on the emulsion and neutron energy. The range of usefulness is limited. The lower limit of measurability is determined by statistical variations and is about 0.1 rem (Q

= 10). The upper limit is set by excessive track density and is about 3 rem (Q = 10). Gamma-ray exposure produces uniform distribution of grains (fog) which obscures the tracks; exposures of a few roentgens prohibit track counting. The neutron response of nuclear track film falls to zero below about 0.5 MeV because there are then fewer than 3 grains per track on the average. From 1 to 10 MeV the number of tracks per unit dose is almost constant. With adequately thick hydrogenous radiators, doses of 20 MeV neutrons can be measured (Becker, 1963).

Further drawbacks of the nuclear-track method are the tedious evaluation procedure (which may be alleviated by the use of a projection microscope) and the fading of nuclear tracks during the time between exposure and processing, so that monitoring periods should be less than two weeks. Fading increases with humidity and temperature and varies with emulsion-type (Zelac, 1968). The use of hermetically-sealed films is recommended for humid environments.

Ordinary film emulsions may be used for thermal neutron monitoring by simple densitometry. It is usual to differentiate thermal neutron dose from gamma-ray dose by comparing the optical densities behind tin and cadmium filters of equal mass per unit area. The cadmium provides additional density through the emission of gamma rays following neutron capture. Dose equivalents of between 50 mrem and 500 rem can be measured provided the gamma-ray background is not too high. The difference in optical density behind the two filters resulting from exposure to 1 rem of slow neutrons is about the same as the optical density that would result behind both filters from exposure to 2 R of radium gamma rays.

Special emulsions loaded with boron or lithium are useful for thermal neutron monitoring. The tracks produced by the alpha particles resulting from (n,α) reactions enhance the sensitivity to thermal neutrons by a factor of 200 compared with unloaded emulsions.

Thermoluminescence. For neutron measurements, LiF phosphor is available using lithium enriched in ^{6}Li, in which the thermal neutron response is enhanced. The response to gamma rays and other radiations may be determined separately by using lithium enriched in ^{7}Li, which exhibits a low thermal neutron response. Thus, the dose from thermal neutrons in a mixed field can be determined by using twin dosimeters with different ^{6}Li content and subtracting one reading from the other. Various schemes using moderators have been used to detect fast neutrons with these phosphors (Dennis *et al.*, 1967).

Radiophotoluminescence. Silver-activated phosphate glass loaded with ^{6}Li is available and may be used to measure thermal neutrons at dose equivalent levels greater than 10 mrem.

Fission Track Devices. Fission fragments cause structural damage along their paths in certain materials which, after a suitable etching process, can be observed as tracks when viewed with optical microscopes. The materials exhibiting this effect include mica, thin foils of plastics (such as Lexan), phosphate glass, and soda lime glass. They can be etched with NaOH, EDTA (ethylenediamine-tetraacetic acid), or HF to enhance track visibility, and examined with a microscope (Fleischer *et al.*, 1965).

Fission damage tracks can be used to detect neutrons and measure neutron fluence (Becker, 1966b). Fissionable material, such as uranium, may be placed against the detecting material or may be evaporated onto it. All of the fission fragments striking the material register tracks. This method can also be used to determine neutron spectra by employing threshold reactions. Depleted uranium and neptunium, with neutron-threshold energies of 1.5 MeV and 0.6 MeV respectively, and plutonium shielded with a thermal neutron shield, with a threshold energy of about 10 keV, may be used (Kerr and Strickler, 1966).

The method has a high sensitivity and is insensitive to gamma rays below the photo-fission threshold (5.2 MeV). The lowest dose of fast neutrons that can be measured with the glass and natural uranium combination is about 10 mrad or 1 mrem. Thermal neutrons can be measured with pure fissile material down to fluence levels of $2 \times 10^4/\mathrm{cm}^2$ or 20 μrem. The detector material will store tracks for more than a year at room temperature and can be reused after simple annealing. Another advantage is the non-destructive readout. One of the most useful materials for fission-track counting is radiophotoluminescent glass (Becker, 1966b). Each individual sample can be used both as a gamma dosimeter by measuring fluorescence, and as a neutron dosimeter by counting tracks.

The disadvantages of the system are the need for employing fissionable material, the highly temperature-sensitive wet etching process, and the care needed to avoid the recording of alpha particles.

Calibration. Most instruments for measuring kerma (rate) from fast neutrons are calibrated by adjusting them to read correctly in a known neutron field. The known fields are produced by sources of known emission rate. One type of such source is made by mixing a radionuclide such as plutonium, polonium, or radium with a material such as beryllium or boron. The neutrons are produced in (α,n) reactions in the latter materials. Radium sources are difficult to use because they also emit intense gamma radiation. The emission rates of these sources are determined by comparison with a source whose emission rate has been measured by absolute methods. Sources of this latter type are kept by the National Bureau of Standards and by some Department of Energy laboratories. One way of comparing emission rates is to use the "precision long counter" (DePangher and Nichols, 1966). This device permits the comparison of sources of different spectra, if enough is known about the spectra. In particular, the emission from targets bombarded by accelerators can sometimes be compared in this way in order to permit using the accelerator for neutron calibrations.

When the emission of the source is known, the flux density at *nearby* points can be calculated from the inverse square law (it is necessary to establish experimentally that the emission from the source is isotropic, or, if it is not, to determine an experimental correction factor). Tables giving the kerma per unit fluence (which is the same as the kerma rate per unit flux density) are given in NCRP Report 25 (NCRP, 1961a) as a function of neutron energy. These factors must be averaged over the spectrum being used. For example, a Pu-Be neutron source produces 3.9 erg/g per 10^7n/cm^2.

The two-ionization chamber systems used for fast neutron measurements in mixed fields of neutrons and gamma rays can also be calibrated with gamma rays only, because there is a fixed relation between the neutron and gamma-ray sensitivities.

The polyethylene proportional counter can also be calibrated with an alpha-particle source. The alpha-particle source is collimated and counted inside the counter behind an electrically operated shutter. While the shutter is open, the alpha particles produce pulses which have identical heights. The energy corresponding to this pulse height can be calculated from the gas pressure, the path length of the particles in the counter, and range data for alpha particles.

The calibrations of rem meters usually are adjusted to indicate the dose equivalent that would be produced in a person at the position of the meter. The calibration therefore allows for neutron scattering within the body and for gamma-ray production *by neutrons* within the body, although the rem meters are themselves insensitive to gamma rays. The calibration does not allow for any incident gamma radiation or for scattering of any radiation from the surroundings into the body. Such gamma radiation *should* be measured separately. Because of the allowance for gamma-ray production within the body, the response of rem meters will be too high for neutrons of less than about 0.1 MeV energy. In most neutron fields from reactors or accelerators, such neutrons produce such a small fraction of the total dose that the error is negligible.

Thermal neutron detectors are usually calibrated by surrounding a fast-neutron source with moderating material to produce thermal neutrons. The flux density of the latter can be determined by absolute activation measurements or by comparison with standard thermal-neutron flux densities at the National Bureau of Standards by use of some stable instrument as an intermediate. It is necessary to establish that placing the instrument to be calibrated in the thermal-neutron field does not alter that field, or to correct the calibration result if it does. This can be done by comparing the readings of a separate very small detector (e.g., an activation foil which does not alter the field) in the presence and in the absence of the instrument to be calibrated.

5.2 Instruments Used for Measurement of Radioactive Material

In radiation protection programs, measurements of activity are required in the determination of body or organ burdens, and in determining activity in the environment, in air, water, food, effluents, and on exposed surfaces. Three general types of instruments are used: instruments for the detection of the presence of radioactive material at a location, instruments for the measurement of activity in a sample and instruments for measurements of activity *in vivo*. For further information refer to NCRP report No. 58 (NCRP, 1978).

5.2.1 *Detection Instruments*

This section covers the various instruments used to detect the presence of radioactive materials on work surfaces, skin, floors, source containers, clothing, in wounds, or on any possibly contaminated surface. Section 3.2.5 outlines the procedures in which instruments are used either for area surveys or for counting wipes and Section 4.4 notes their application to personnel contamination monitoring.

5.2.1.1 *General Instrument Properties.* Monitoring by means of wipes, instrument surveys of benchtops, clothing, skin, and other surfaces is at best a qualitative procedure and it is difficult to make it highly quantitative. Therefore, the instruments used are usually of the detector type rather than of the measuring type. However, since the amount of activity involved is usually small, the sensitivity of the instruments *should* be high.

The necessity for portability of contamination detectors depends on their intended uses. If the instrument is for general purpose monitoring of laboratory surfaces, a portable type is required. If the instrument is for a specific use in which the item to be monitored can be brought to the instrument, portability is not required. Clothing and hand and shoe monitors are generally not portable.

Count-rate instruments usually incorporate meter readouts. Portable count-rate meters usually have earphone jacks, and bench monitors often have aural outputs. Specialized monitors can have numerical readout and background subtracting capabilities. Table 2 summarizes the various kinds of contamination-detection instrumentation.

Careful calibration of these instruments is not required. It is usually sufficient to test their operation and overall sensitivity with small sealed check sources.

5.2.1.2 *Alpha Contamination Detectors.* Instruments used for the detection of alpha contamination will record all alpha particles entering the sensitive volume. The sensitivity of these instruments is therefore determined by the area of the window. The area of the detector is commonly about 50 cm^2 and the window thickness 1 mg/cm^2. Alpha contamination monitors *should* be insensitive to beta and gamma radiation in order to reduce the background reading. This is usually accomplished by pulse height discrimination in the counting circuit.

Portable alpha monitors are available that use gas proportional counters or scintillation counters. Proportional counter rate meters use either air or special gas mixtures as the counting gas. The scintillation counter probes use silver-activated ZnS phosphor. These detectors are relatively insensitive to gamma rays and exhibit an extremely low background. Care *should* be exercised in field use to prevent puncturing the window. In a scintillation counter, a damaged window may produce a light leak that can "jam" the counter.

Certain end-window G-M counters have windows thin enough to admit alpha particles, but are also sensitive to beta and gamma radiations and therefore exhibit a background counting rate. The detection of a small amount of alpha contamination with a G-M counter is difficult because the increase in counting rate above the background level is very small.

The probes used in the portable instruments can be used with some line-operated bench-monitor circuits, provided the amplification and high voltage supply match the probe characteristics.

5.2.1.3 *Beta Contamination Detectors.* Portable counters of several types can be used for the detection of beta-particle contamination of surfaces. G-M count-rate meters used for this purpose require thin end-windows or walls of thin metal or drawn glass (thickness of 30–40 mg/cm^2). Scintillation counters can also be used for beta contamination detection. Thin sections of anthracene or plastic scintillator coupled to a photomultiplier are very sensitive to beta particles and relatively insensitive to gamma rays. A light shield consisting of a thin opaque window is necessary in this type of counter. If the contaminant

TABLE 2—*Contamination detection instruments*

Instrument	Range of Counting Rate and Other Characteristics	Typical Uses	Remarks
Beta-Gamma Surface Monitors[a]			
General			
Portable Count Rate Meter (Thin Walled or Thin Window G-M Counter)	0–1,000; 0–10,000; 0–1000,000 count/min	Surfaces, hands, clothing	Simple, reliable; battery powered
Thin End Window G-M Counter Lab Monitor	0–1,000; 0–10,000; 0–100,000 count/min	Surfaces, hands, clothing	Line-operated
Personnel			
Hand and Shoe Monitor G-M or Scintillation Counter Type	From 1-½ to 2 times natural background	Rapid contamination monitoring	Automatic operation
Special			
Laundry Monitors, Floor Monitors, Doorway Monitors	From 1-½ to 2 times natural background		Convenient and rapid
Alpha Surface Monitors			
General			
Portable Air Proportional Counter With Probe	0–100,000 dis/min over 100 cm^2	Surfaces, hands, clothing	Not applicable in high humidity; battery powered; fragile window
Portable Gas Flow Counter With Probe	0–100,000 count/min over 100 cm^2	Surfaces, hands, clothing	Not affected by humidity; battery powered; fragile window
Portable Scintillation Counter With Probe	0–100,000 count/min over 100 cm^2	Surfaces, hands, clothing	Not affected by humidity; battery powered; fragile window
Personnel			
Hand and Shoe Monitor Proportional Counter Type	0–2,000 count/min over about 300 cm^2	Rapid contamination monitoring of hands and shoes	Automatic operation
Hand and Shoe Monitor Scintillation Type	0–4,000 count/min over about 300 cm^2	Rapid contamination monitoring of hands and shoes	Rugged
Wound Monitors	Low energy photon detection	Monitoring for plutonium	Not commercially available
Air Monitors			
Particle Samplers			
Filter Paper High Volume	40 ft^3/min (1.1 m^3/min)	For quick grab samples	Used intermittently; requires separate counter
Filter Paper	0.1 to 10 ft^3/min (0.003–0.3 m^3/min)	For continuous room air breathing zone monitoring	Used continuously; requires separate counter
Electrostatic Precipitator	3 ft^3/min (0.09 m^3/min)	For continuous monitoring	Sample deposited on cylindrical shell, requires separate counter
Impinger	20 to 40 ft^3/min (0.6–1.1 m^3/min)	Alpha contamination	Special uses; requires separate counter
Tritium Monitors			
Flow ionization chambers	0–10 $\mu Ci/m^3$ /min	Continuous monitoring	May be sensitive to other sources of ionization
Complete Monitoring Systems			
Fixed Filter Paper	Minimum detectable activity: $\beta\gamma$: ~ 10^{-12} $\mu Ci/cm^3$ α: ~ 10^{-12} $\mu Ci/cm^3$		Background buildup can mask low level activity; counter included
Moving Filter Paper	Minimum detectable activity: $\beta\gamma$: ~ 10^{-12} $\mu Ci/cm^3$ α: ~ 10^{-12} $\mu Ci/cm^3$		Continuous record of air activity. Time of measurement can be adjusted from time of collection to any later time

[a] None of these surface monitors is suitable for tritium detection.

is a low-energy beta emitter such as carbon-14, sulfur-35, or calcium-45, a thin end-window G-M counter is necessary. End-window counters are available with 1 mg/cm^2 windows.

Portable counters generally *cannot* be used for detecting tritium because the beta-particle energy is too low (E_{max} = 18 keV) to allow the particle to enter the counter. Wipes are the usual means of surveying for tritium contamination of surfaces. To assay such wipes, counters capable of detecting very low-energy beta particles such as liquid scintillation counters or windowless gas-flow counters (discussed in Section 5.2.3) *should* be used. A paper strip used for wiping or alternatively a drop of solvent (chosen for the chemical compound being sought) from a cotton wipe can be added to a counting vial for liquid scintillation counting. Detection of airborne tritium is discussed in Section 5.2.2.1.

The beta-gamma detectors described above can be connected to a line-operated rate meter or to bench-monitor circuits if portability is not necessary. Automatic instruments, such as hand and shoe monitors, laundry monitors, floor monitors, and doorway monitors are available for the detection of beta contamination under special circumstances. The use of these instruments is limited to situations where many repetitive measurements of the same type are necessary. In laboratories having only a few workers, a bench monitor with a thin end-window G-M counter connected by a long cable can be used as a beta-particle hand and shoe monitor.

Instruments used for beta-particle or gamma-ray measurements will respond to background radiation, which must be taken into account in the interpretation of the reading. If a high background radiation level exists, the usefulness of counters for contamination monitoring is limited since they do not indicate small increases in initially high counting rates. Under these conditions wipes are recommended (see Section 3.2.5).

5.2.1.4 *Gamma Contamination Detectors.* Since most gamma-ray emitters also emit beta particles, most contamination monitors are made to detect both radiations. The beta-ray detectors described in Section 5.2.1.3 can be used in gamma-ray contamination monitors with the exception that thin windows or walls are not necessary. The usual practice is to use detectors sensitive to both radiations in order to increase sensitivity, since the detection efficiency is usually greater for beta particles than for gamma rays. Sometimes contaminants are covered by enough material to totally absorb the beta particles, or pure gamma emitters are involved. Plastic scintillators or NaI crystals are more sensitive to gamma rays than G-M counters and are therefore recommended for detecting gamma rays.

5.2.2 *Instrumentation for Air and Liquid Monitoring*

Instruments used for the monitoring of contaminants in air or effluents *should* have high sensitivity because the amount of radioactive material involved is usually small. Since the measurements are frequently only qualitative, the instruments need not be highly accurate.

5.2.2.1 *Air Samplers and Monitors.* Particulate contamination in the air can be monitored by pumping the air through a filter and measuring the activity retained on the filter. Specific types of samplers include "grab" samplers, continuous monitors with fixed filter paper, and continuous monitors with moving filter paper.

"Grab" samples are taken with an air suction pump containing a filter-paper holder. They can be used for rapid assays after radioactive spills and after environmental air contamination by radioactive particulates. A flow meter to measure the amount of air pumped is required for quantitative results. Often the radionuclide constituting the contaminant is known and the filter paper is assayed for its particular radiation. In order to obtain a reasonable sample in a short time, an air pump capable of pumping about 10 to 40 ft^3/min (0.3–1.1 m^3/min) is generally used, and sampling times are of the order of 0.1 to 2 hours.

If routine air monitoring is required in an area in which there is only minor contamination, a sampler can be operated continuously and the sample counted at specific intervals, e.g., once per day. An example of this type of monitor consists of an air pump of 1 to 10 ft^3/min (0.03–0.3 m^3/min) capacity, a flow meter, and a filter paper holder. Similar devices, provided with an intake that can be located close to the nose of a worker, and frequently portable, are used to monitor the breathing zone. Flow rates of the order of 1 ft^3/min (0.03 m^3/min) are typical.

Monitoring may also be done with air monitors that count the sample at the time of collection, if the information is desired immediately. In this type of monitor, an appropriate counter is placed near the filter paper and an indication of any change in the activity of the collected dust is obtained. Since the smallest amount of activity that can be detected above background is relatively large, this type of monitor is most useful in situations where large air pollution incidents may occur. A continuous air monitor *should* be equipped with an alarm since its principal function is to warn the worker. Several of these complete continuous air-monitoring systems are available commercially.

For measurement of alpha-particle contamination, a surface-loading filter (such as a membrane filter) must be used and the sample collected must be thin. The usefulness of air monitors for detecting alpha-particles from a particular nuclide can be increased by using semiconductor detectors and a single-channel analyzer. The resolution of surface barrier type detectors permits discrimination against alpha particles from radon and its daughters, while the alpha particles from a specific radionuclide are counted. Air monitoring for plutonium or uranium can be selectively performed in this manner.

The moving-filter-paper air monitors use a strip of filter paper that slowly passes first over the collection point and then in front of the detector or detectors. A delayed, continuous record of the airborne activity is thus obtained. The time between collection and measurement can be varied to allow for decay of the daughter products of radon. The filter paper moves with a speed of about an inch per hour.

In monitoring for radioactive gases a concentrated sample is difficult to obtain. Systems designed specifically for some gases are available but generally are of low sensitivity. Iodine-131 can be detected with filter paper (particularly if loaded with charcoal or silver nitrate) because some of the iodine deposits on the paper, but quantitative measurements require a chemical process or an activated charcoal trap providing efficient absorption.

Filter-paper air samplers can be calibrated by counting the filter paper and a standard source in the same geometrical arrangement, and correcting for the collection efficiency of the filter taking account of the size of particles collected and air velocity employed.

Tritium monitoring. Tritiated water vapor usually exists in the vicinity of nuclear reactors and nuclear fuel reprocessing plants. Tritium is used as a target material in some neutron generators (Section 3.3.4.5) and is used extensively in biological and physical research often at levels of 1 curie or more. Tritium gas following release into the atmosphere slowly converts at lower altitudes to tritiated water and mixes with environmental water. Hence, tritiated water and tritium gas are the primary forms of tritium contamination. NCRP Report No. 47, *Tritium Measurement Techniques* (NCRP, 1976a), discusses their collection and measurement in detail. Budnitz (1974a) has recently reviewed instrumentation for tritium monitoring.

Tritiated water has some affinity for dust, but the use of filter paper techniques for air sampling described above will not collect all of the airborne tritiated water and will result in low estimates of its concentration. The most sensitive and accurate measurement techniques involve the absorption or condensation of tritiated water vapor. Condensation is achieved by passing a sample of air through a cold trap that uses an alcohol-dry ice bath (Chiswell and Dancer, 1969) or liquid nitrogen (Iyengar *et al.*, 1965). The condensate is preferably counted in a liquid scintillation counter (Section 5.2.3.5). The relative humidity and air temperature must be measured to determine the effective air volume sampled. Concentrations of 0.01 percent of the occupational MPC can be readily measured by this method. Absorption of tritiated water vapor from an air sample is accomplished by passing the sample through a trap containing a drying agent (Osloond *et al.*, 1973) or a molecular sieve (Chiswell and Dancer, 1969) or by bubbling it through distilled water (Banville, 1965; McConnon, 1970). Detection of concentrations less than 10^{-5} MPC is reported with techniques utilizing silica gel as a drying agent and direct counting of the gel in a liquid scintillation counter (Osloond *et al.*, 1973).

Tritium in the air (e.g., as hydrogen, hydrocarbons, or water vapor) can be measured rapidly with Kanne chambers, that are ionization chambers through which the air flows (Reinig and Albanesius, 1962). The method is only marginally able to detect levels below the occupational MPC when portable chambers with volumes of about 1 liter are used. In chambers with volumes of the order of 10 liters (usually non-portable), 0.1 MPC can be measured. Since the detector is an ionization chamber, external sources of radiation affect the reading. To avoid this difficulty, two identical chambers can be employed, one for sampling and one for measuring the background, and the difference in the ionization currents measured (Osborne and Coveart, 1973). If problems arise through a response due to air ionization by agents other than radiation, an electrostatic deionizer can be installed in the air stream before the ionization chamber. The deionizer will remove all ions in the entering air, including those created by tritium, and therefore will greatly reduce the sensitivity of the measuring chamber, but readings will be quantitative.

Radon monitoring. Radon gas appears in the atmosphere near any deposit of natural uranium. Significant concentrations occur near uranium mill tailings, while, in uranium mines, exposure to radon daughter products constitutes an important hazard. This hazard, together with the unique radiation characteristics of radon and its daughters, has prompted the development of specialized monitoring instruments, particularly for use in mines. The instrumentation has been critically reviewed by Budnitz (1974b) and is discussed in American National Standard N13.8 (ANSI, 1973).

Several classes of monitoring instruments have been and are being developed that measure: (a) radon gas concentration; (b) "Working Level"; or (c) radon daughter concentrations. "Working Level" is a dosimetric concept similar to a maximum permissible concentration relating the concentration of radon daughters in air to the integrated alpha particle energy emission. One Working Level is equivalent to the energy release of the short-lived daughters in equilibrium with 100 pCi of radon in 1 liter of air and is defined as: any combination of short-lived radon progeny in one liter of air that will result in the ultimate emission of 1.3×10^5 MeV of alpha energy during decay to lead-210.

Typical of the pure radon measuring instruments is the *Lucas chamber* (Lucas, 1957) that comprises a chamber of some 200 ml volume coated internally with zinc sulfide and into which radon is drawn after filtration. The filter removes the radon daughters. Scintillations due to the radon alpha particles are counted with a photomultiplier tube which views the walls of the chamber through a clear flat window. The sensitivity of the method is about 10 pCi/liter. Another common method that also measures radon concentrations immediately but is more sensitive than the Lucas chamber is the *two-filter method.* Sample air is pumped through a metal cylinder about 1 meter long and 10 cm in diameter which is fitted with a filter on each end. The input filter removes the radon daughters. Some of the RaA generated by radon decay in the cylinder is deposited on the exit filter and its alpha emission is immediately counted with a ZnS scintillation counter. Sensitivities of the order 1 pCi/liter are obtained (Thomas and LeClare, 1970). Both methods are adequate for routine use in mines.

In other methods the daughter products are counted after reaching equilibrium (in about 3 hours), at which time the count-rate is proportional to the radon concentration. In one instrument the alpha rays from RaA and RaC′ deposited on a fine membrane filter of less than 1 μm pore size are counted. In another instrument filtered radon is trapped in an activated charcoal filter at dry ice temperature and the gamma rays from RaC in the charcoal are counted with a scintillation counter after equilibrium is reached. This method can detect about 0.1 pCi/l of radon.

The standard method for Working Level measurement is that of Kusnetz (1956) in which the daughter products RaA, RaB, and RaC from a known gas volume are collected on a filter and the RaA plus RaC′ alpha activity is subsequently counted after a fixed delay. The alpha count rate can be related to the integrated alpha energy release from these daughters.

A common method for measurement of individual daughter concentrations is that of Tsivoglou *et al.* (1953). Again, an air sample of about 50 liters is passed through a filter but subsequently the alpha activity is measured at three later times, usually 5, 15, and 30 minutes after collection. From these data the original concentrations of RaA, RaB, and RaC can be determined. An alternative method involving only two counts at different times employs alpha spectroscopy of the 6.00 MeV and 7.69 MeV alpha particles from RaA and RaC′, respectively. As developed by Martz *et al.* (1969), the method employs a solid state detector and multichannel analyzer for spectral analysis.

5.2.2.2 *Liquid Samplers.* Continuous monitoring of water and waste lines for radioactive materials is sometimes necessary. Waste lines from hot laboratories and reactor cooling lines are examples. Complex equipment can be used to monitor the line continuously or the liquid may be sampled and the sample analyzed in the laboratory (Straub, 1964).

For continuous beta-gamma contamination monitoring, several systems are available in which G-M dip counters or scintillation counters immersed in the flow, or jacketed G-M counters, are connected to rate meters and recorders. The sensitivity of these systems is relatively low, since the detector is exposed to the radioactivity of a limited volume. Hence, they are useful in monitoring liquids into which large quantities of radioactive material may be accidentally released. The presence of low energy beta-particle emitters such as tritium cannot be detected with this type of monitor. Contamination may build up on the detector and may cause a variable or gradually increasing background giving misleading readings.

Continuous monitoring can be performed, however, by routine laboratory analysis of a small sample proportional to the flow rate. Samplers that take a periodic aliquot or that continuously extract a small amount of the liquid are available. The samples are then prepared and analyzed as discussed in Section 5.2.3.

In small installations, liquid monitoring is usually performed on periodic samples, that are prepared and analyzed in the laboratory. This is the usual method of determining the radioactive material concentration in a hold-up tank in order to derive the permissible discharge rate.

5.2.3 *Instruments for the Measurement of Activity in Samples*

Samples requiring routine laboratory assay for radionuclide content include air, water, urine, gas, and various solid samples such as vegetation and ground samples and wipes. Often the radionuclide is known and the analysis is confined to a particular radiation. If the nuclide is not known, the sample *should* be examined for the presence of both alpha and beta-gamma emitters. Radiochemical separation for particular nuclides is not discussed here.

Most samples analyzed in radiation protection programs contain very small amounts of activity. The analyzing instrumentation *should* be highly sensitive and the effect of natural background radiation levels on the detectors *should* be reduced. Background can be adequately reduced by the use of shielded counting rooms or local shields around the detectors. Large installations routinely analyzing many samples frequently use both types of shielding. A review of sample-counter requirements is given in Table 3. Detailed accounts of techniques for the measurement of low-level radioactivity in samples are contained in ICRU Report 22 (ICRU, 1972) and NCRP Report No. 50 (NCRP, 1976).

5.2.3.1 *Sample Preparation.* Preparation of the various samples requiring analysis for radioactive material content may necessitate reducing the sample to a form suitable for counting and concentrating the radioactive material into a small volume to increase the sensitivity. Samples requiring no preparation prior to counting include airborne particulates on filter papers from air samplers and wipes, which may be placed in a planchet and counted directly. Small, weighed, soil samples or other solid samples may also be counted directly for assaying gamma emitters.

Water samples to be analyzed for gross activity are prepared by evaporating a known volume of a liter or less to dryness. The water can be slowly boiled in a beaker to a small volume of 10 to 50 ml, transferred to a planchet, and evaporated to dryness under an infrared lamp. Detection of gamma-ray emitters in water samples can be performed without sample preparation by using well counters. Liquid samples to be analyzed for specific nuclides, such as some urine samples, require complex chemical separations. In the analysis of

TABLE 3—*Sample counter requirements*

Counter Type	Preamplifier	Amplifier Gain	System Sensitivity	High Voltage Supply	Sample Sensitivity (μCi)[c]
G-M	Not required	Not required	0.1 V	500–1500	beta 10^{-4} gamma 10^{-2}
Proportional Gas Flow Counter	Desirable[a]	Gain of 1000	1 mV	500–2500 (low noise)	beta 10^{-5}
Gamma Scintillation Counter	Desirable[a,b]	Gain of 300	50 mV	500–1500 (low noise)	well: 5×10^{-5} probe: 10^{-2}
Liquid Scintillation Counter	Desirable[b]	Gain of 3000	5 mV	500–2500 (low noise)	beta 10^{-5}
Alpha Scintillation Counter	Not required	Not required	0.1V	500–1500	5×10^{-4}
Semi-Conductor Detector	Charge-Sensitive Preamp Required	Pulse-Shaping Amplifier Required	—	0–500 (low noise)	alpha < 1 dpm gamma 5×10^{-5}

[a] If cable between detector and amplifier is long (greater than 5 feet) preamplifier may be required.

[b] Preamplifier is required for pulse height analyzer input.

[c] Generally background-equivalent activity.

urine, the naturally occurring potassium-40 *should* be chemically separated from the sample to reduce the natural background.

Solid samples *should* be prepared so as to make them compatible with the counting equipment. Accurate measurements of alpha or beta emitters in solid samples will usually require special sample preparation such as dry ashing for organic materials or electrochemical deposition after dissolution. A review of the preparation of thin alpha-emitting sources has been given by Yaffe (1962). For alpha or beta counting the counter *should* be calibrated with sources physically similar to the expected samples in order to reduce counting inaccuracies due to backscatter, self absorption of the sample, and air absorption.

5.2.3.2 *Ionization Chambers.* Ionization chambers are used for the measurement of radioactivity in gases. The concentration of tritiated water vapor or of radioactive noble gases in air can be assayed by use of a sealed ionization chamber and an electrometer. An ionization chamber of known volume is filled with the radioactive gas to a known pressure and either the ionization current or the charge collected in a given time is measured. Typical measuring systems employ ionization chambers of approximately one liter in volume and vibrating-reed electrometers. Compared with counting techniques (e.g., liquid scintillation counters for tritium measurements), the sensitivity of ionization chambers is poorer by several orders of magnitude, and therefore their use has been superseded largely by proportional and liquid scintillation counters.

Special chambers used for tritium measurements are discussed in Section 5.2.2.1. Similar gas flow counters are available for monitoring noble gases (e.g., xenon-133) in room air at the level of the occupational MPC.

5.2.3.3 *Geiger-Mueller Counters.* A common arrangement for the measurement of beta and gamma activities consists of a lead enclosure containing a thin end-window G-M counter connected to a scaler and a sample holder with shelves at various distances from the window. Maximum permissible concentrations in air and water for beta-gamma emitters can be measured in reasonable times with this detector by use of normal sampling procedures and preparation, and maximum permissible contamination levels can be detected on wipes.

Background counting rates of the order of 30 counts per minute from this type of counter can be reduced by the use of flat G-M counters, which have approximately the same sensitivity to the radiations from the sample as the normal tubular type. Single electronic units containing a scaler, timer, and high voltage supply are available for use with G-M counters. The scaler *should* have an input pulse sensitivity of about 0.1 V, a counting capacity of 5 or more decades, and provision for presetting the counting time. The high voltage supply *should* be adjustable to 1500 V.

The overall efficiency of the counting system *should* be determined with standard sources for various shelves in the sample holder when the counter is set up, and at regular intervals thereafter. Typical overall efficiencies for the shelf closest to the counter lie between 25 and 35 percent for beta-particles above 100 keV.

Windowless G-M counters can be used to measure low-energy beta radiation. These counters operate at a pressure slightly above atmospheric with a constant flow of special counting gas passing through the counter. The sample is inserted into the sensitive volume and the counter sealed. The geometrical efficiency is close to 50 percent. Conventional scalers as described above can be used with these counters.

For the analysis of special samples, special shapes and arrangements of G-M counters can be used. For instance, cylindrical electrostatic precipitation samples can be measured using a thin-wall G-M counter mounted in the center of the sample.

5.2.3.4 *Proportional Counters.* Proportional counters for sample evaluation offer the advantages over G-M counters of discrimination between different types of radiation and ability to count at much higher rates. Since alpha particles produce much larger pulses than beta particles or gamma rays, they can be counted separately by proper electronic discrimination. Proportional counters are widely used with thin windows, or window-less, especially for counting samples containing low-energy beta-ray emitters. They are operated as flow counters at atmospheric pressure for counting solid samples. Radioactive noble gases, $^{14}CO_2$ and tritium, can be counted following their introduction into the counting gas in a closed system. The counting gas is often methane or argon/methane in a 90/10 ratio. For counting very low-level samples, a guard counter may be placed around the main counter and anti-coincidence circuitry used to discriminate against background events registering in both counters. With this system, the background count-rate can be reduced to less than 1 cpm (ICRU, 1972).

Internal gas counting is the most sensitive available technique for assaying the concentration of radioactive noble gases. For example, ^{85}Kr can be measured at its present ambient level of about 10 pCi/m^3 of air (Jaquish and Moghissi, 1973). The use of larger chamber volumes operated at several atmospheres pressure with rise-time discrimination enhances the sensitivity (Bradley and Willes, 1972).

Tritium in the form of hydrogen or hydrocarbons but preferably not tritiated water may be counted with great sensitivity in proportional counters of the closed or gas-flow type (Cameron, 1967). In a 1-minute count, activity equivalent to about 1 pCi/cm^3 of a water sample can be detected. This sensitivity is greater than that of liquid scintillation counters largely because of the lower background counting rate.

5.2.3.5 *Scintillation Counters.* Samples may be assayed for alpha, beta, and gamma activity with scintillation counters. Alpha counters employ thin coatings of ZnS phosphors on photomultiplier-tube faces. The sample must be positioned close to the detector. The phosphor may also be deposited on a nylon film which is placed in contact with the sample and discarded after use. The efficiency of this type of counter is determined with calibrated alpha sources and approaches 50 percent. The background counting rate is extremely low.

Beta-ray counters usually employ thin sections of plastic scintillators or thin anthracene crystals optically coupled to a photomultiplier tube. The efficiency for energetic beta particles approaches 100 percent and is low for high-energy gamma radiation.

Particular geometric configurations of scintillation counters can be used to assay samples for gamma emitters. Large volume NaI(T1) or plastic scintillator well counters can be used to measure the gamma-ray activity in liquid samples by inserting a flask of the liquid directly in the well. Calibration of the counter *should* be performed by utilizing a physically similar standardized source. Standards prepared from mixed solutions of gamma-ray emitters available, for example, from the National Bureau of Standards, are particularly convenient since they permit determination of the curve of detector efficiency versus energy in a single measurement.

The use of NaI scintillators permits the identification of gamma-ray emitters in samples if pulse height analysis is performed. Long counting times may be necessary for low-activity samples. If enough activity is present, the activity of a given radionuclide can be determined by computing the area under the photo-peak. The availability of large NaI scintillators [3 to 9 inches (7.6 to 22.9 cm) in diameter] surrounded by massive shields has increased the use and accuracy of this technique.

For counting low-energy beta-particle emitters such as tritium, carbon-14, sulfur-35, and calcium-45, the severe loss of counts through self-absorption in solid samples, and the small size of the pulses relative to the noise pulses in solid scintillation counters, have made liquid scintillation counters employing coincidence counting the method of choice. The filter paper used in air monitors or for wiping can be introduced directly into the liquid scintillator. Liquid from a moist cotton ball used for wiping can also be counted in this manner.

For counting organic samples, toluene or p-xylene are the most common solvents for the scintillation fluor, which is usually either PPO (diphenyloxazole) or p-terphenyl. In connection with radiation protection measurements, the sample to be counted is frequently aqueous (e.g., urine) and such samples may greatly reduce (quench) the light output from the above solution. Such quenching is minimized by adding alkylphenol detergents (Lieberman and Moghissi, 1970) that permit counting in a solution containing water up to 50 percent by volume. Alternatively, a mixture of p-dioxane as a primary solvent with naphthalene as a secondary solvent is widely used and can accommodate water up to 20 percent without appreciable quenching (Moghissi *et al.*, 1969).

Coincidence counting of the pulses generated by two photomultiplier tubes is a standard technique in liquid scintillation counters. Older counters were generally refrigerated to about 0°C (273°K) in order to reduce the background counting rate due to random coincidences. In modern counters, using photomultipliers with bi-alkali cathodes and high-speed electronics that allow coincidence gating times of the order of 20 nsec., adequately low random coincidence (less than 10 cpm) can be obtained without refrigeration. In these counters, typical efficiencies for unquenched samples are 60 percent for tritium and close to 100 percent for carbon-14.

Calibration of liquid scintillation counters for particular radionuclides is achieved by adding to particular samples a known standard activity of the radionuclide in a form that will not affect the degree of quenching. The increased count-rate divided by the disintegration rate of the standard represents the counting efficiency. Variations in quenching from sample to sample may be determined by observing the additional count-rate produced by a gamma-ray emitting standard external to the sample container.

5.2.3.6 *Semiconductor Counters.* Semiconductor detectors may be used for counting samples containing alpha-particle, beta-particle, or gamma-ray emitters. Two main types are presently available: the surface-barrier silicon p-n junction detector, which is used to detect alpha particles and beta particles; and the lithium-drifted silicon or germanium p-i-n junction detector, which is used predominantly for gamma-ray spectroscopy.

The semiconductor radiation detector may be regarded as an ionization chamber with a solid detecting medium. Solids are about 1000 times as dense as gases and therefore much smaller thicknesses are needed to absorb radiation. While charged particle spectrometry with gaseous detectors is confined to those producing short, densely-ionized tracks, such as alpha particles, semiconductors can be used for a much wider range of radiation types. Moreover, for a given amount of absorbed energy, semiconductors produce about 10 times as many elementary charge carriers. The reduction in statistical fluctuations substantially improves the energy resolution. In comparison with gas counters, semiconductor detectors are also smaller, have much shorter rise times (of the order 1 nanosecond), show greater long-term stability, and are free of gas purity problems. On the other hand, the signal is typically small so that more sophisticated amplifiers are required, and there is a higher level of system noise that determines the lower energy detection limit.

The silicon surface barrier detector for alpha particle and beta particle measurement is a diode consisting of a very thin p-type layer on the face of a high purity n-type silicon wafer. Radiation is admitted through a thin gold film on the p-type surface about 40 $\mu g/cm^2$ thick to which electrical contact is made. The sensitive volume (charge depletion region) increases with the bias voltage applied and the resistivity of the n-type material and may range up to a few millimeters. A beta particle of energy 1.2 MeV has a range of about 2 mm in silicon. Detectors with areas up to 10 cm^2 are available.

The high resolution of these detectors for alpha particles (typically about 10 keV) permits the identification of the components of mixtures of alpha emitters. Samples can be analyzed quantitatively if the counter has been calibrated with sources whose activity is known. In order to realize the high resolution possible, the sample and detector *should* be in a vacuum chamber to reduce air absorption and scattering.

For gamma-ray spectroscopy, germanium detectors are preferred to silicon because of the much higher photoelectric absorption efficiency (about 40 times greater). High efficiency also requires a large sensitive volume or depletion thickness, which in turn requires high resistivity (intrinsic) material in the depleted region. This is usually achieved by compensating the acceptor impurities in p-type material by drifting lithium, a donor impurity, into it; alternatively, ultra-high-purity germanium can be used. Depletion thicknesses of 1 cm or more are readily obtainable and sensitive volumes of 50 cm^3 are common in coaxially drifted detectors (IAEA, 1966). The maintenance of lithium compensation requires that lithium-drifted detectors be kept at very low temperature, but this is not necessary with pure germanium detectors.

The great advantage of these gamma ray detectors is their superior energy resolution, which makes them uniquely suitable for quantitative analysis of mixtures of gamma-ray emitters. For 1 MeV photons, resolution (FWHM) of about 2 keV is possible, compared to about 80 keV with NaI(T1) scintillation counters. For 30 keV photons resolution of about 300 eV is common. Such resolution requires operation of both the detector and the preamplifer at liquid nitrogen temperature (77°K).

Calibration of pulse height in terms of photon energy is performed with a series of x- and gamma-ray emitters or with mixed radionuclide sources. For quantitative analysis, spectra of the pure radionuclides are required (ICRU, 1972).

5.2.3.7 *Photographic Film.* Photographic film is used in sample analysis by autoradiography. For example, small amounts of alpha activity found in processed urine samples can be measured by placing nuclear-track film in contact with the electrodeposited, chemically separated emitter for extended periods of time and then counting the nuclear tracks in the developed emulsion. The advantage of this system is the extremely high sensitivity that can be realized because the exposure time is usually a few weeks. The disadvantages are the long exposure times necessary and the elaborate film processing and reading techniques.

5.2.4 *Radioactive Material in the Human Body*

Gamma-ray emitting radionuclides within the body can be detected from outside the body with highly sensitive gamma-ray instruments (IAEA, 1962; Meneely and Line, 1965; IAEA, 1970a). One of the principal problems is to distinguish between the gamma radiation from the body and that from the surroundings. The human body normally contains about 10^{-13} Ci/g of gamma emitters. Most common materials in the surroundings contain about 10^{-12} Ci/g. Thus, measurements of normal activity levels in the body require the use of good shielding to absorb rays coming from the surroundings. Clean iron shields contain about 10^{-14} Ci/g, and facilitate the measurement of normal activity levels in the body.

Special shielding is not required to measure maximum permissible body burdens (MPBB) of most beta-gamma emitting radionuclides that are about 1 μCi or higher, i.e., about 10^{-11} Ci/g. However, much smaller body burdens must be measured to maintain a control program that will prevent the MPBB values from being reached. As a consequence, most measuring systems require some shielding.

5.2.4.1 *Whole-Body Counters.* Two principal types of whole-body counters are in use. One type uses one or more large NaI(Tl) scintillation crystals to detect the gamma rays. A typical size of crystal is 20 cm in diameter and 10 cm in thickness. Counting periods as long as 5 to 30 minutes are required to measure body burdens at the normal level. A great advantage of these counters is the fact that pulse height analysis can be used to determine the energy of the gamma rays detected, a procedure that permits identification of the radionuclides present. Shielding against background radiation is usually accomplished by housing the counter and the subject in a small iron room with walls 6 to 10 inches (15.2–25.4 cm) thick. Talc (McCall, 1960) and chalk (Trott *et al.*, 1963) have also been used for shielding.

The other principal type of whole-body counter employs plastic or liquid scintillators. Large volume scintillators can be used, and these can achieve high sensitivity. It is possible, for example, to surround the subject with an annular liquid scintillator. With this arrangement counting times of only a few minutes are adequate for measurement at background levels. Indeed, if a grossly contaminated subject is encountered, the sensitivity may prove so great that the counting equipment will "jam". Pulse-height analysis is not normally feasible with plastic or liquid scintillators, but may be possible if the potential contaminants are known. The shielding requirements for these counters are slightly more modest than for NaI counters, because more counts can be gathered in a short time, an advantage permitting a statistically more accurate subtraction of background.

Either type of whole-body counter can measure body burdens of gamma emitters as low as 1 nCi. The body normally contains natural potassium-40 equivalent to about 10 nCi of a gamma emitter with 100 percent gamma-ray abundance, and about 10 nCi of cesium-137 from fallout.

Transportable whole-body counters have been developed (Naversten *et al.*, 1963; Brady and Swanberg, 1965; Boddy, 1966; Parker and Anderson, 1967). These usually employ a single NaI crystal, but have lighter shields. Either the volume shielded is smaller, or use is made of the "shadow shield" principle (Palmer and Roesch, 1965). In shadow shielding, the main shield is placed immediately around the detector and its opening is directed toward a second shield beyond the subject that is large enough to intercept background radiation from this direction. The sensitivities of these counters are a little lower than those discussed above. These counters have been installed in trucks so that measurements can be made at work locations, or at the scene of accidents, or at places convenient for persons living near a facility (Howard *et al.*, 1971).

5.2.4.2 *Other Instruments for the Detection of Gamma Emitters.* Conventional scintillation counters are used to measure radioiodine in the thyroid, sodium-24 produced by neutron irradiation, and other radioactive materials in the body. These are usually counters employing NaI(Tl) crystals with dimensions of a few inches. Thyroid radioiodine burdens of a few nanocuries can be measured by one or two scintillation counters placed close to the gland (Laurer and Eisenbud, 1963). Accuracy can be improved by corrections for gland depth (Wellman *et al.*, 1967). Thin-crystal detectors are preferable for the measurement of the low-energy radiation from iodine-125 and are commercially available. Portable G-M counter survey instruments have been used to detect neutron-produced sodium-24 (Wilson, 1962).

5.2.4.3 *Detection of Beta Emitters.* From outside the body it is very difficult to detect a radionuclide that emits no gamma rays. All beta-particle emitters produce some x rays during the emission and slowing down of the beta particle. These can be detected, but this has not proved useful in practice because it has not been possible to distinguish them from internal gamma rays scattered by the body.

However, a system for measuring phosphorus-32 has been developed (Palmer, 1966) that utilizes two counters operated in coincidence to reduce background. The beta rays emitted through the skin of the face pass through a thin, transmission proportional counter, and are then detected in a broad, thin NaI(Tl) scintillation counter with a thin entrance window.

5.2.4.4 *Detection of Alpha Emitters.* Alpha emitters in the body are detectable only by virtue of the gamma rays or x rays that are also emitted. Radionuclides such as radium, thorium, and uranium, which emit gamma rays, require whole-body counters for their measurement, even at maximum permissible levels, because the MPBB's for these alpha emitters are much lower than those for beta-gamma emitters. The gamma rays from uranium are of low energy and in addition require special techniques (Cofield, 1960).

Plutonium-239 emits almost no gamma rays. However, 4 percent of the disintegrations emit 13 to 20 keV x rays produced in internal-conversion processes. Such low-energy x rays are very easily absorbed in the body. Thin, broad NaI(Tl) scintillation counters with thin entrance windows are very sensitive to these x rays. In this way, it is possible to detect as little as 0.1 nCi of plutonium in shallow wounds from which no alpha particles can be detected (Roesch and Baum, 1958; Epstein and Johanson, 1966; Cloutier and Watson, 1967). Internally deposited plutonium, e.g., that located in the lung, liver, or bone, is much harder to detect. Special large-area scintillation and proportional counters are available that will detect much less than the maximum permissible lung burden of 16 nCi for ^{239}Pu.

Proportional counters using xenon gas filling can achieve backgrounds of less than 1 cpm and detection limits of a few nanocuries using anticoincidence methods (Ramsden, 1969; Ramsden, 1976). Ishihara *et al.* (1969) described an 8 inch diameter, 5 mm thick sodium iodide detector that will detect 6 nCi in the lung for a 100 minute count. Swinth and Griffin (1970) have developed a multicrystal unit with 1 mm thick crystals covering a total area of 6 in. by 11 in. for lung counting. More recently the new principle of "phoswich detectors" has been developed in which a sodium iodide-cesium iodide pair of scintillators is employed discriminating against background counts due to Compton scatter by means of pulse shape discrimination (Laurer and Eisenbud, 1968; Tomlinson, 1976; Dean *et al.*, 1970). Detectability of 9 nCi in a 30 minute count is reported by Voelz *et al.* (1976).

The principal problem in all these devices is the difficulty of calibration for lung burden because of the strong absorption of the soft x rays in overlying tissue (HVL-0.6 cm of tissue), almost complete absorption in ribs, and the uncertainty of distribution of plutonium.

6. Radiation Accident Monitoring

6.1 General

As the result of a radiation accident, an emergency situation may occur that could involve:

(1) High dose rates and/or contamination levels at the site of the event;
(2) Highly contaminated, exposed, and possibly injured people; or
(3) Serious consequences extending far beyond the normal radiation control area.

All facilities where such accidents are possible *shall* have equipment and procedures that will permit the following prompt actions:

(1) Identification of persons who have received large radiation doses;
(2) Identification of persons who are highly contaminated or injured;
(3) Initiation of actions for treatment;
(4) Evaluation of the dose received by persons involved in the accident;
(5) Measurement and control of the dose to personnel during the accomplishment of necessary tasks immediately after the accident (such as rescue, treatment, and stabilization of the event);
(6) Execution of an off-site, as well as on-site, radiological emergency plan.

This section is primarily addressed to the instrumentation and monitoring methods required for accomplishing the above objectives. Dosimetry for radiation accidents is reviewed extensively in the proceedings of two conferences convened by the International Atomic Energy Agency (IAEA, 1965; 1970b).

6.1.1 *Radiation Accident*

A radiation accident may be defined as an unpredictable occurrence, involving exposure to radiation or contamination of humans or the contamination of the environs with radioactive materials. The discussion in this section will be limited to accidents involving exposure greatly in excess of permissible limits and/or the spread of contamination in sufficient quantities to restrict seriously or to terminate the operation of the facility.

Radiation accidents may be caused by failure of mechanical or electrical equipment, by accidental or deliberate damage to equipment containing or intended to contain radioactive material, or by disregard of safety precautions. The actual radiation accident involving exposure to individuals or the spread of radioactive materials normally will encompass only a short span of time. The dose received by individuals at the time of occurrence of the accident is termed an "accident dose".

6.1.2 *Emergency Exposures*

As a result of a radiation accident, high radiation or contamination levels may exist in a location where prompt action must be taken to avoid personal exposure and spread of the contamination. In some unusual circumstances, this action may require that selected personnel be exposed above the permissible limits. To the extent possible, a single emergency dose *should* be limited to 25 rad to the whole body. Individuals who have in the past received such emergency doses *shall not* again be exposed above permissible limits.

6.1.3 *Special Needs for Accident Monitoring*

Instruments *should* be provided for the following purposes:

(1) Detection of the accident.
(2) Survey of the accident area, possibly by remote control methods.
(3) Monitoring of involved persons for contamination and during decontamination.
(4) Monitoring of personnel engaged in evacuation or rescue.
(5) Determination of dose received by persons involved in the accident.

Radiation monitoring equipment that is useful in the detection of a radiation accident and the notification of personnel of its occurrence can vary from highly sophisticated detection and alarm equipment to normal radiation monitoring equipment discussed in Section 5. In some instances the equipment that indicates a possible radiation accident may not be radiation detection equipment. For example, an alarm from a fire detection system in a facility where large quantities of radioactive materials are used *should* be an indication that a radiation accident may have occurred.

The identification of personnel involved in the accident and the determination of their contamination and dose status may require the use of various combinations of fixed area monitors, personnel dosimeters, contamination monitoring, whole-body counting, and assay of biological samples. The role of each technique is discussed below. The methods for measurement and control of exposure of personnel who must re-enter the affected location are also discussed.

6.2 Area Survey Under Accident Conditions

An essential part of coping with any radiation accident is the prompt determination of the radiation status at the event site and in surrounding areas. Changing conditions *should* be detected quickly to prevent involvement of large numbers of people and large quantities of equipment.

Surveys of the event site with portable instruments *should* be made to provide basic data on the radiation levels present. These *should* be planned carefully to limit the exposure of emergency personnel. Although the preparation and methods for making the survey are basically the same as those described in Section 3, the following unusual circumstances *should* be considered:

(1) The location of sources of radiation may be unknown;
(2) Physical safeguards may have been destroyed;
(3) The physical process or reaction that caused the accident may still be occurring; and
(4) A criticality reaction could be triggered again by the approach of the individual making the survey.

The worst possible conditions *should* be assumed. At least two high range instruments having capability to detect alpha, beta, and gamma radiations (neutrons if possibly present) *should* be checked for proper operation prior to entry. High range direct reading and/or alarm dosimeters *should* also be checked and set.

Careful attention *shall* be given to the safety of the surveyor, both radiological and physical, as described in Section 3.2.3. The "buddy" system *should* be adopted for all entries into the affected area, principally to assure the physical safety of the personnel conducting the survey. However, the number of instruments and measurements required and the need for a rapid but thorough appraisal of the conditions within the accident site also dictate the need for more than one individual per team.

The survey *should* be designed to obtain gross answers concerning the status of the facility. Precise answers are not required immediately and may never be required. In order to conserve time, no attempt *should* be made to correct instrument readings. This refinement can be made at a later time based on the data accumulated from the survey and the instrument capabilities.

For some accidents, a gamma camera has proved useful for identification of localized high-level radiation sources in a given area. The camera functions similarly to a pinhole camera but utilizes lead shielding. Both photographic and radioautographic film are included in the film cassette to provide a photograph as an overlay to the radiographic reproduction. Techniques such as use of the attenuation of the surveyor's body or other objects to assist in locating the radiation source(s) are useful.

After the radiation levels have been determined, the magnitude and extent of the contamination spread *should* be established by a rapid survey. The survey may entail the measurement of contamination levels directly on equipment and surfaces or it may require the collection of wipes for evaluation outside of the event site.

6.3 Personnel Monitoring

6.3.1 *Determination of Accident Dose*

The rapid assessment of dose from external sources is important in the determination of medical treatment required for individuals involved in a radiation accident. The level and type of external or internal contamination *should* also be rapidly determined since it may present a greater hazard to involved personnel than the dose received from external radiation fields. If, for example, there has been an intake of certain bone-seeking radionuclides, it may be advisable to initiate medical treatment within an hour of the time of the accident.

Personnel dosimeters are necessary to assist in the evaluation of radiation dose received by individuals at the time of the event or during any rescue, recovery, or reclamation programs. These dosimeters *should* be collected, surveyed for contamination, and processed immediately. Special care *should* be taken to assure that all elements of the dosimeter remain identified with the individual during the processing. For example, after an accident, an identification number on a film carried by an individual may be obliterated because of high level dose received or special processing techniques used. In these situations alternative methods of positive identification of the dosimeter *should* be used, such as a perforation code.

Fixed area dosimeters may be used to provide information in case personnel dosimeters are not worn or are lost. Sometimes a more accurate measurement may be made by a fixed area dosimeter than by a dosimeter worn on the human body. In general, however, personnel dosimeter results are preferred. Fixed area dosimeters *should* be collected as soon as radiation levels and emergency procedures permit. Special care *should* be taken to assure that all elements of these dosimeters remain identified with a specific location during the processing.

In a criticality accident, a large part of the dose received by involved personnel may be due to neutrons. Appendix F of NCRP Report No. 38 (NCRP, 1971a) contains a detailed discussion of accident dosimetry with emphasis on the estimation of neutron dose.

If personnel contamination is not present, the neutron dose received may be approximately established by a "quick-sort" method based on measurements of sodium-24 activity induced in body fluids (Wilson, 1962). A G-M type monitoring instrument is placed against the abdomen of the exposed individual. The activation level per rad of neutron dose depends on the neutron spectrum and may vary by a factor of 3. As a rough guide, after whole body exposure to an unmoderated fission spectrum, the increase in count-rate for one rad of neutron dose will be approximately equal to the normal background.

More precise evaluation of neutron doses of the order of 10 rad may be accomplished by determining the sodium-24 activity in blood. Approximately 10 ml samples of blood (100 ml samples if the dose is less than 10 rad) treated with an anti-coagulant *should* be analyzed according to established procedures to determine the amount of sodium-24 present. A conversion factor for unmoderated fission spectrum neutrons is 1.66×10^5 rad per μCi/ml of sodium-24 in the blood (Union Carbide, 1958; Hurst *et al.,* 1959; Smith, 1962). This factor, as a function of time after exposure, is shown in Figure 3.

If the relationship between the fluence of neutrons above 2.9 MeV and the total neutron fluence is known, the neutron dose to localized areas or critical organs of the body can be estimated by determining the phosphorus-32 activity resulting from the (n,p) activation of sulfur-32 in hair (Petersen *et al.,* 1961). Samples of one gram of hair from head, chest, groin, legs, and back areas *should* be analyzed separately for phosphorus-32 according to established analytical procedures. The fluence of neutrons with energies above 2.9 MeV (n/cm^2) may be

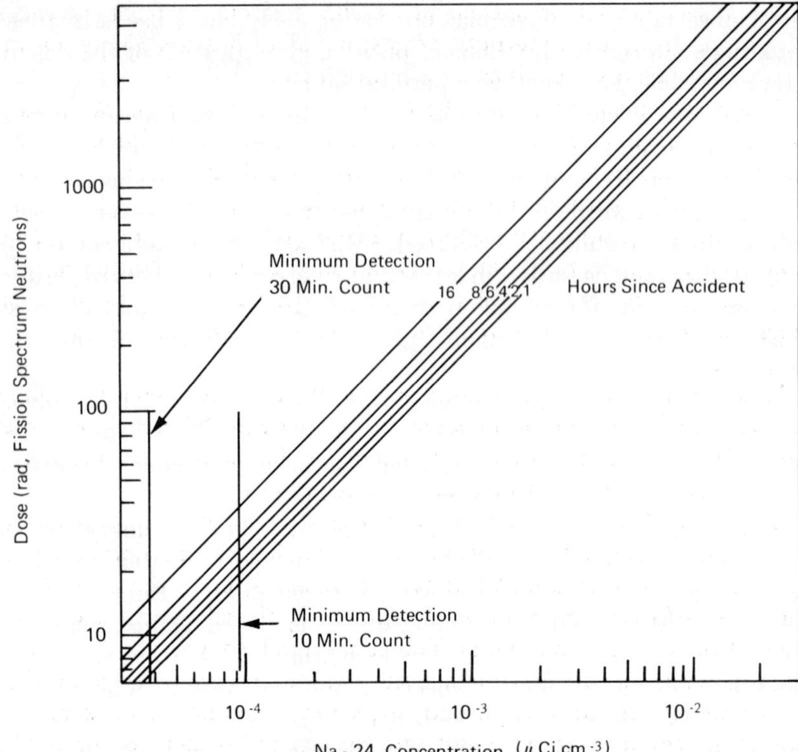

Fig. 3. Neutron dose versus concentration of sodium-24 in blood.

calculated as follows (Hankins, 1969)[18]:

$$F = \frac{AW}{\sigma s N \lambda m} = 1.46 \times 10^8 \frac{A}{m} \quad (1)$$

F = fast neutron fluence detected by sulfur (n/cm^2)
A = activity of phosphoros-32 (dis/min at t = 0)
W = atomic weight of sulfur (32)
σ = activation cross-section of sulfur (240×10^{-27} cm^2/atom)
s = sulfur content per gram of hair (0.0448)
N = Avogadro's number (6.02×10^{23})
λ = decay constant (3.38×10^{-5}/min)
m = weight of hair sample (g)

[18] Fluence to dose conversion for neutrons with energies in excess of 2.9 MeV may be assumed to be 4.4×10^{-9} rad per neutron/cm^2 for a critical assembly. For an unmoderated fission spectrum, neutrons above the sulfur threshold contribute approximately 30 percent to the dose from neutrons.

The analysis of radioactivity induced in metallic items to determine neutron fluence may provide valuable data in the absence of other sources or may confirm measurements from other sources. The estimation of dose *should* be made in a manner similar to the evaluation of dosimeter foils.

As mentioned previously, it is possible to determine the sodium-24 content by performing *in vivo* measurements. If a whole-body counter facility is available, a rapid determination of fast-neutron dose from a fission spectrum may be obtained from the following relationship (Larson and Keene, 1965):

$$D_s = \frac{Q_s}{w} (2.15 \times 10^2) \tag{2}$$

where D_s = average total-body fast-neutron dose in rad
Q_s = sodium-24 activity in μCi
w = weight of subject in kilograms.

Special measurements may be required to assess the amount of contamination present in a wound or deposited internally; e.g., radionuclides in specific organs such as iodine-131 in the thyroid and plutonium-239 in wounds and in lungs (See Section 5.2.4).

6.3.2 *Estimation of Dose During Emergency Monitoring*

Measurement of dose received by rescue and monitoring personnel during an emergency *should* be made. However, it is necessary to estimate, during the course of these activities, the dose that is likely to be received. To permit this rapid estimation of projected emergency dose, special signaling dosimeters, direct reading dosimeters, or integrating dose rate monitoring devices with a range of at least twice the expected or planned emergency dose *should* be available for each emergency team member. Dosimeters are available in several ranges, from zero to 10, 25, 50, 100, or 600 rad.

The exposures received during emergency work, as determined by regular dosimeters, special devices mentioned above, or dose-rate time measurements *should* be recorded.

6.4 Radiation Accident Instrumentation

Types of instruments together with some of their characteristics, that would satisfy radiation monitoring requirements, are discussed below. Instruments discussed in Section 5 are not included in this discussion.

6.4.1 *Accident Detection Instruments*

Area surveillance equipment, such as a high-level gamma detector, that is to be used in the detection of an accident or in the determination of radiation field levels or contamination levels after a radiation accident, *should* be of a type that does not become saturated and that exhibits stable performance over long periods. The equipment *should* be designed to provide a positive response to any malfunction. The exposure range for this equipment may vary from a few mR/h to 100 R/h and the equipment may contain an adjustable alarm point. Criticality detectors, in addition to having most of the characteristics listed above, *should* be capable of responding to pulsed radiation, whose burst duration is as low as 50 microseconds. Both types of instruments *should* have an independent power supply and a unique alarm.

Waste stream and air contamination monitors may be useful in providing early warning of an impending accident in addition to providing valuable data after the accident. A sensitivity of 10^{-6} μCi/cm^3 for the waste stream monitor and 10 MPC-h for the air contamination monitor would be desirable. In any event, the monitors *should* be capable of detecting 10 times the allowable release limit in one hour.

6.4.2 *Dose-Rate Survey Instruments*

If an accident that would cause extremely high dose rates can occur, special high dose-rate instruments *should* be provided and kept available for immediate use. Instruments that are capable of measuring exposure rates of up to 100 R/min with an accuracy of ± 25 percent are available. In some cases, the detector can be located several feet from the instrument by use of rigid or cable extensions. The correct functioning of these instruments *should* be checked quarterly.

6.4.3 *Contamination Survey Instruments*

Specialized contamination survey instruments are available for use in determining the levels of contamination over surface areas of several square meters. These utilize various plastic or solid state detectors and the appropriate electronics to permit the measurement of radioactive materials in quantities greater than 1 μCi/m^2.

6.4.4 *Personnel Dosimeters*

Several types of personnel dosimeters are available that provide satisfactory measurements of x or gamma rays. A commonly used personnel dosimeter is the film badge. Normally the film packet contains either two films or two types of emulsions on the same film base, one sensitive to x- and gamma-ray exposures in the range of 10 mR and the other sensitive to a few R up to 500 R when conventional methods of film processing and film-density measurements are employed. Teflon discs containing thermoluminescent ^{7}LiF and ^{6}LiF to provide an additional indication of neutron dose may also be used (Nash and Attix, 1971). Some personnel dosimeters contain a system of activation foils (see Section 5.1.2.4) (Adams, 1970). Measurements with these foils permit the estimation of neutron dose up to 2000 rad for energies ranging from thermal to 15 MeV. Generally, nuclear track film will not be suitable for dose measurement after radiation accidents because the associated gamma radiation may be so high as to darken the film completely and prevent neutron track counting.

6.4.5 *Fixed Area Dosimeters*

Fixed area dosimeters (Reinig *et al.,* 1965; Bramson, 1963) *should* include gamma and neutron capabilities as required to serve emergency needs of the installation. Gamma dosimeters are available with ranges up to 10,000 R for gamma energies of 10 keV to 10 MeV. Thermoluminescent dosimeters have practical advantages over chemical and polyacrylamide dosimeters (Duffy and Kasper, 1968). Neutron dose measurements may be made with any of a number of dosimeter foil systems for neutron energies from thermal to 15 MeV and doses up to 2000 rad. Foils may be used as activation foils or as threshold detector foils. Commonly used threshold detectors and their reactions are listed in Section 5.1.2.4.

References

ACGIH (1971). American Conference of Governmental Industrial Hygienists, *Threshold Limit Values of Airborne Contaminants and Physical Agents with Intended Changes Adopted by ACGIH for 1971* (American Conference of Governmental Industrial Hygienists, Cincinnati).

ADAMS, N. (1970). "Review of United Kingdom research and experience in criticality dosimetry," page 79 in *Nuclear Accident Dosimetry Systems,* IAEA Publication No. STI/PUB/241 (International Atomic Energy Agency, Vienna).

AEC (1970). U.S. Atomic Energy Commission, *Incineration of Radioactive Solid Wastes,* Report No. WASH-1168 (National Technical Information Service, Springfield, Virginia).

ANDERSSON, I. O. (1961). "A proton-recoil proportional counter for fast neutron dosimetry," page 453 in *Selected Topics in Radiation Dosimetry,* IAEA Publication No. STI/PUB/25 (International Atomic Energy Agency, Vienna).

ANDERSSON, I. O. AND BRAUN, J. (1964). "A neutron rem counter," Nukleonik **6,** 237.

ANSI (1966). American National Standards Institute, *American National Standard Practice for Occupational Radiation Exposure Records System,* ANSI Report No. N13.6-1966 (R1972) (American National Standards Institute, New York).

ANSI (1969a). American National Standards Institute, *Guide for Administrative Practices in Radiation Monitoring,* ANSI Report No. N13.2-1969 (American National Standards Institute, New York).

ANSI (1969b). American National Standards Institute, *Guide to Sampling Airborne Radioactive Materials in Nuclear Facilities,* ANSI Report No. N13.1-1969 (American National Standards Institute, New York).

ANSI (1969c). American National Standards Institute, *Radiological Safety in the Design and Operation of Particle Accelerators,* ANSI Report No. N43.1-1969, published as National Bureau of Standards Handbook No. 107 U.S. Government Printing Office, Washington.

ANSI (1971). American National Standards Institute, *Marking Physical Hazards and the Identification of Certain Equipment (Safety Color Code),* ANSI Report No. Z53.1-1971 (American National Standards Institute, New York).

ANSI (1972a). American National Standards Institute, *Performance Specifications for Direct Reading and Indirect Reading Pocket Dosimeters for X- and Gamma Radiation,* ANSI Report No. N13.5-1972 (American National Standards Institute, New York).

ANSI (1972b). American National Standards Institute, *Radiation Safety for X-Ray Diffraction and Fluorescence Analysis Equipment,* ANSI Report No. N43.2-1971, published as National Bureau of Standards Handbook No. 111 (U.S. Government Printing Office, Washington).

ANSI (1973). American National Standards Institute, *Radiation Protection in Uranium Mines,* ANSI Report No. N13.8-1973 (American National Standards Institute, New York).

ANSI (1975). American National Standards Institute, *General Safety Standard for Installations Using Non-Medical X-Ray and Sealed Gamma-Ray Sources, Energies up to 10 MeV,* ANSI Report No. N543, published as National Bureau of Standards Handbook No. 114 (U.S. Government Printing Office, Washington).

ATTIX, F. H. (1972). "A current look at TLD in personnel monitoring," Health Phys. **22,** 287.

AXTON, E. J. AND BARDELL, A. G. (1972). "Neutron production from electron accelerators used for medical purposes," Phys. Med. Biol. **17,** 293.

BANVILLE, B. (1965). *Calibration of Tritium-in-Air Monitor,* Report No. AECL-2265 (Atomic Energy of Canada, Ltd., Chalk River Nuclear Laboratories, Ontario).

BARBIER, M. (1969). *Induced Radioactivity* (John Wiley and Sons, New York).

BEACH, S. A. AND DOLPHIN, G. W. (1964). "Determination of plutonium body burdens from measurements of daily urine excretion," page 603 in *Assessment of Radioactivity in Man,* Vol. II, IAEA Publication No. STI/PUB/84 (International Atomic Energy Agency, Vienna).

BECKER, K. (1963). "Source of error in neutron personnel dosimetry with nuclear track emulsion," Atomkernenergie **8,** 74.

BECKER, K. (1966a). *Photographic Film Dosimetry* (The Focal Press, New York).

BECKER, K. (1966b). "Nuclear track registration in dosimeter glasses for neutron dosimetry in mixed radiation fields," Health Phys. **12,** 769.

BECKER, K. (1973). *Long-Term Stability of Film, TLD, and Other Integrating Dosimeters in Warm and Humid Climates,* Report No. ORNL-TM-4297 (Oak Ridge National Laboratory, Oak Ridge, Tennessee).

BELCHER, E. H. AND GEILINGER, J. E. (1957). "Improved scintillating media for radiation dosimetry," Br. J. Radiol. **30,** 103.

BERGER, M. J. AND SELTZER, S. M. (1970). "Bremsstrahlung and photoneutrons from thick tungsten and tantalum targets," Phys. Rev. C **2,** 621.

BLUM, M. AND LIUZZI, A. (1967). "Thyroid ^{131}I burdens in medical and paramedical personnel," J. Am. Med. Assoc. **200,** 992.

BODDY, K. (1966). "A mobile shadow-shield whole-body monitor for medical research," Health Phys. **12,** 1148.

BRADLEY, A. E. AND WILLES, E. H. (1972). *Risetime Discrimination for Low Level Tritium Counting,* Report No. UCRL 73978 (Lawrence Livermore Laboratory, Livermore, California).

BRADLEY, F. J. AND JONES, A. H. (1970). "RF response of radiation survey instruments," page 173 in *Electronic Product Radiation and the Health*

Physicist, DHEW Publication No. BRH/DEP 70-26 (Bureau of Radiological Health, Rockville, Maryland).

BRADY, D. N. AND SWANBERG, F., Jr. (1965). "The Hanford mobile whole-body counter," Health Phys. **11,** 1221.

BRAMBLETT, R. L., EWING, R. I. AND BONNER, T. W. (1960). "A new type of neutron spectrometer," Nucl. Instrum. Methods **9,** 1.

BRAMSON, P. E. (1963). *Hanford Criticality Dosimeter,* Document No. HW-71710 (Revised) (General Electric Company, Hanford Atomic Products Operation, Richland, Washington).

BRH (1974). Bureau of Radiological Health, *Suggested State Regulations for Control of Radiation* (Available from Bureau of Radiological Health (HFX-25), Food and Drug Administration, Rockville, Maryland).

BROBECK, W. M. AND ASSOCIATES (1968). *Particle Accelerator Safety Manual,* Appendices B and C, Report No. MORP 68-12 (National Center for Radiological Health, Rockville, Maryland).

BROOKE, C. AND SCHAYES, R. (1966). "Recent developments in thermoluminescent dosimetry; extensions in the range of applications," page 31 in *Solid State and Chemical Radiation Dosimetry in Medicine and Biology,* IAEA Publication No. STI/PUB/138 (International Atomic Energy Agency, Vienna).

BRYNJOLFSSON, A. AND MARTIN, T. G., III (1971). "Bremsstrahlung production and shielding of static and linear accelerators below 50 MeV. Toxic gas production, required exhaust rates and radiation protection instrumentation," Int. J. Appl. Radiat. Isot. **22,** 29.

BUCHAN, R. C. T. (1968). "The limitations of personal radiation monitoring in the radiodiagnostic department," Br. J. Radiol. **41,** 876.

BUDNITZ, R. J. (1974a). "Tritium instrumentation for environmental and occupational monitoring—a review," Health Phys. **26,** 165.

BUDNITZ, R. J. (1974b). "Radon-222 and its daughters—a review of instrumentation for occupational and environmental monitoring," Health Phys. **26,** 145.

BUSH, D. AND HUNDAL, R. S. (1973). "The fate of radioactive materials burnt in an institutional incinerator," Health Phys. **24,** 564.

BUSHONG, S. C., HARLE, T. S. AND WRIGHT, A. E. (1969). "Personnel radiation monitoring in the radiodiagnostic department," Br. J. Radiol. **42,** 632.

CAMERON, J. F. (1967). "Survey of systems for concentration and low background counting of tritium in water," page 543 in *Radioactive Dating and Methods of Low-Level Counting,* IAEA Publication No. STI/PUB/152 (International Atomic Energy Agency, Vienna).

CARUTHERS, L. T. AND MAXWELL, R. B. (1971). "Contamination of a university print shop from the cleaning of a static eliminator device," Health Phys. **21,** 713.

CFR (1976a). Code of Federal Regulations, Title 21, Subchapter J, Part 1020.10, *Performance Standard for Television Receivers* (U.S. Government Printing Office, Washington).

CFR (1976b). Code of Federal Regulations, Title 21, Subchapter J, Part 1020, *Performance Standard for Ionizing Radiation Emitting Products* (U.S. Government Printing Office, Washington).

CFR (1976c). Code of Federal Regulations, Title 21, Subchapter J, Part 1020, Section 1020.40, *Performance Studies for Cabinet X-Ray Systems* (U.S. Government Printing Office, Washington).

CFR (1976d). Code of Federal Regulations, Title 10, Part 20, *Standards for Protection Against Radiation* (U.S. Government Printing Office, Washington).

CFR (1976e). Code of Federal Regulations, Title 21, Subchapter J, Subpart C—*Radiation Protection Recommendations* (U.S. Government Printing Office, Washington).

CHEKA, J. S. (1954). "Recent developments in film monitoring of fast neutrons," Nucleonics **12,** No. 6, 40.

CHISWELL, W. D. AND DANCER, G. H. C. (1969). "Measurement of tritium concentration in exhaled water vapour as a means of estimating body burdens," Health Phys. **17,** 331.

CHISWELL, W. D. AND GILBOY, W. B. (1972). "Comparison of the radiation doses to the wrists and fingers of workers engaged on radiochemical processing," Health Phys. **22,** 327.

CIPPERLY, F. V. (1966). "Thermoluminescent dosimetry monitoring system for the National Reactor Testing Station," Health Phys. **12,** 1637.

CLOUTIER, R. J. AND WATSON, E. E. (1967). "Measuring plutonium in wounds: advantages of using the americium gamma ray," Health Phys. **13,** 811.

COFIELD, R. E. (1960). "*In vivo* gamma counting as a measurement of uranium in the human lung," Health Phys. **2,** 269.

COOK, C. S. (1963). *Residual Radioactivity Following Cyclotron Shutdown,* Report No. TR-667 (U.S. Naval Radiological Defense Laboratory, San Francisco).

COX, F. M. AND LUCAS, A. C. (1974). "An automated thermoluminescent dosimetry (TLD) system for personnel monitoring: (Part 1) system description and preliminary results," Health Phys. **27,** 339.

DADE, M., HOEGL, A. AND MAUSHART, R. (1972). "RPL dosimetry system with fully automated data evaluation," page 693 in *Proceedings of the 3rd International Conference on Luminescent Dosimetry, Riso,* Report No. Riso-249 (Part 2) (Danish Atomic Energy Commission, Riso, Denmark).

DEAN, P. N., IDE, H. M. AND LANGHAM, W. H. (1970). *External Measurement of Plutonium Lung Burdens,* LASL Report No. LA-DC-10906 (Los Alamos Scientific Laboratory, Los Alamos, New Mexico).

DENNIS, J. A. AND LOOSEMORE, W. R. (1961). "A fast-neutron counter for dosimetry," page 443 in *Selected Topics in Radiation Dosimetry,* IAEA Publication No. STI/PUB/25 (International Atomic Energy Agency, Vienna).

DENNIS, J. A., SMITH, J. W. AND BOOT, S. J. (1967). "Measurement of the dose-equivalent from thermal and intermediate-energy neutrons with personnel dosimeters," page 537 in *Neutron Monitoring,* IAEA Publication No. STI/PUB/136 (International Atomic Energy Agency, Vienna).

DEPANGHER, J. (1959). "Double moderator neutron dosimeter," Nucl. Instrum. Methods **5,** 61 or Report No. HW-57293 (General Electric Company, Hanford Atomic Products Operation, Richland, Washington).

DEPANGHER, J. AND NICHOLS, L. L. (1966). *Precision Long Counter for*

Measuring Fast Neutron Flux Density, Report No. BNWL-260 (Battelle Pacific Northwest Laboratories, Richland, Washington).

DHEW (1965). U.S. Department of Health, Education, and Welfare, *Procedures for the Determination of Stable Elements and Radionuclides in Environmental Samples,* U.S. Public Health Service, Publication No. 999-RH-10 (U.S. Department of Health, Education, and Welfare, U.S. Public Health Service, Washington).

DHEW (1967). U.S. Department of Health, Education, and Welfare, *Radioassay Procedures for Environmental Samples,* U.S. Public Health Service, Publication No. 999-RH-27 (U.S. Department of Health, Education, and Welfare, U.S. Public Health Service, Washington).

DHEW (1968). U.S. Department of Health, Education, and Welfare, *Technical Papers: Conference on Detection and Measurement of X-Radiation from Color Television Receivers, March, 1968* (U.S. Department of Health, Education, and Welfare, National Center for Radiological Health, Bureau of Radiological Health, Rockville, Maryland).

DHEW (1970a). U.S. Department of Health, Education, and Welfare, *Product Testing and Evaluation of Giant View TV Projector,* Technical Report No. BRH/DEP 70-9 (U.S. Department of Health, Education, and Welfare, Bureau of Radiological Health, Rockville, Maryland).

DHEW (1970b). U.S. Department of Health, Education, and Welfare, *Laboratory Testing and Evaluation of Color Television Receivers Acquired During the In-plant Survey,* Technical Report No. BRH/DEP 70-6 (U.S. Department of Health, Education, and Welfare, Bureau of Radiological Health, Rockville, Maryland).

DHEW (1971). U.S. Department of Health, Education, and Welfare, *Direct-Print Paper: A New Material for X-Ray Imaging,* Electronic Products Technical Note, DHEW Publication No. (FDA) 72-8013, BRH/DEP N72-03 (Bureau of Radiological Health, Rockville, Maryland).

DHEW (1974). U.S. Department of Health, Education, and Welfare, *BRH Routine Compliance Testing for Diagnostic X-Ray Systems,* DHEW Publication No. (FDA) 75-8012 (U.S. Department of Health, Education, and Welfare, Bureau of Radiological Health, Rockville, Maryland).

DUFFY, T. L. AND KASPER, R. B. (1968). "Studies of gamma dosimetry systems used for nuclear accident dosimetry," Health Phys. **14,** 45.

EHRLICH, M. (1954). *Photographic Dosimetry of X- and Gamma Rays,* National Bureau of Standards Handbook No. 57 (National Bureau of Standards, Washington).

Electronic Industries Association (1972). *Recommended Practice on X-Ray Detection and Measurement for Microwave Tubes,* JEDEC Publication No. 70-A (Electronic Industries Association, Washington).

ELS, R. A. (1971). "The response of various survey instruments to low-energy x-rays," page 141 in *Radiation Safety in X-Ray Diffraction and Spectroscopy,* DHEW Publication No. (FDA) 72-8009, BRH/DEP 72-3 (Bureau of Radiological Health, Rockville, Maryland).

EPSTEIN, R. J. AND JOHANSON, E. W. (1966). "Apparatus for monitoring ^{239}Pu in wounds," Health Phys. **12,** 29.

FLEISCHER, R. L., PRICE, P. B. AND WALKER, R. M. (1965). "Tracks of charged particles in solids," Science **149,** 383.

FOWLER, I. L., PEARSON, A. AND WATT, L. A. K. (1959). "Halogen beta-counter monitors for Chalk River reactors," Nucleonics **17,** 78.

GEYER, J. C., MACMURRAY, L. C., TALBOYS, A. P. AND BROWN, H. W. (1956). "Low level radioactive waste disposal," page 19 in *International Conference on Peaceful Uses of Atomic Energy,* Vol. 9, Geneva (United Nations, New York).

GOLLNICK, D. A. (1972). "An inexpensive TLD personnel badge," Health Phys. **23,** 255.

HANKINS, D. E. (1968). "The substitution of a BF_3 probe for the LiI crystal in neutron rem-meters," Health Phys. **14,** 518.

HANKINS, D. E. (1969). "Direct counting of hair samples for ^{32}P activation," Health Phys. **17,** 740.

HARRIS, P. S. (1961). "Acute radiation death resulting from an accidental nuclear critical excursion. Section VI. Dosimetric calculations," J. Occup. Med. **3,** 178.

HEALY, J. W. (1970). *Los Alamos Handbook of Radiation Monitoring,* Report No. LA-4400 (Los Alamos Scientific Laboratory, Los Alamos, New Mexico).

HERZ, R. H. (1969). *The Photographic Action of Ionizing Radiations* (John Wiley and Sons, Inc., New York).

HORNYAK, W. F. (1952). "A fast neutron detector," Rev. Sci. Instrum. **23,** 264.

HOWARD, L. E., JR., SPICKARD, J. H. AND WILHEMSEN, M. (1971). "A human radioactivity counter and medical van," Health Phys. **21,** 417.

HOWLEY, J. R. AND ROBBINS, C. (1967). "Radiation hazards from x-ray diffraction equipment," Radiol. Health Data Rep. **8,** 245 (U.S. Government Printing Office, Washington).

HURST, G. S. (1954). "An absolute tissue dosemeter for fast neutrons," Br. J. Radiol. **27,** 353.

HURST, G. S., RITCHIE, R. H. AND EMERSON, L. C. (1959). "Accidental radiation excursion at the Oak Ridge Y-12 plant—III. Determination of radiation doses," Health Phys. **2,** 121.

IAEA (1962). International Atomic Energy Agency, *Whole-Body Counting,* IAEA Publication No. STI/PUB/47 (International Atomic Energy Agency, Vienna).

IAEA (1965). International Atomic Energy Agency, *Personnel Dosimetry for Radiation Accidents,* IAEA Publication No. STI/PUB/99 (International Atomic Energy Agency, Vienna).

IAEA (1966). International Atomic Energy Agency, *Lithium-Drifted Germanium Detectors* IAEA Publication No. STI/PUB/132 (International Atomic Energy Agency, Vienna).

IAEA (1970a). International Atomic Energy Agency, *Directory of Whole-Body Radioactivity Monitors,* IAEA Publication No. STI/PUB/213 (International Atomic Energy Agency, Vienna).

IAEA (1970b). International Atomic Energy Agency, *Nuclear Accident Dosimetry Systems,* IAEA Publication No. STI/PUB/241 (International Atomic Energy Agency, Vienna).

ICRP (1964). International Commission on Radiological Protection, *Radiation Protection,* ICRP Publication No. 6, paragraph 86 (Pergamon Press, New York).

ICRP (1966). International Commission on Radiological Protection, International Commission on Radiological Protection Task Group on Lung Dynamics, "Deposition and retention models for internal dosimetry of the human respiratory tract," Health Phys. **12,** 173.

ICRP (1968a). International Commission on Radiological Protection, *Evaluation of Radiation Doses to Body Tissues from Internal Contamination due to Occupational Exposure,* ICRP Publication No. 10 (Pergamon Press, New York).

ICRP (1968b). International Commission on Radiological Protection, *The Assessment of Internal Contamination Resulting from Recurrent or Prolonged Uptakes,* ICRP Publication No. 10A (Pergamon Press, New York).

ICRP (1970) International Commission on Radiological Protection, *Protection against Ionizing Radiation from External Source*, ICRP Publication No. 15 (Pergamon Press, New York)

ICRP (1973). International Commission on Radiological Protection, *Implications of Commission Recommendations that Doses be kept as low as Readily Achievable* ICRP Publication No. 22 (Pergamon Press, New York).

ICRP (1975). International Commission on Radiological Protection, "Physiological data for reference man," Chapter 3 in *Report of the Task Group on Reference Man,* ICRP Publication No. 23 (Pergamon Press, New York).

ICRP (1977). International Commission on Radiological Protection, *Recommendations of the International Commission on Radiological Protection*, ICRP Publication No. 26 (Pergamon Press, New York)

ICRU (1963). International Commission on Radiation Units and Measurements, *Radiobiological Dosimetry,* ICRU Report 10e, published as National Bureau of Standards Handbook No. 88 (International Commission on Radiation Units and Measurements, Washington).

ICRU (1964). International Commission on Radiation Units and Measurements, *Physical Aspects of Irradiation,* ICRU Report 10b, published as National Bureau of Standards Handbook No. 85 (International Commission on Radiation Units and Measurements, Washington).

ICRU (1971). International Commission on Radiation Units and Measurements, *Radiation Protection Instrumentation and Its Application,* ICRU Report 20 (International Commission on Radiation Units and Measurements, Washington).

ICRU (1972). International Commission on Radiation Units and Measurements, *Measurement of Low Level Radioactivity,* ICRU Report 22 (International Commission on Radiation Units and Measurements, Washington).

ICRU (1977). International Commission on Radiation Units and Measurements, *Neutron Dosimetry for Biology and Medicine,* ICRU Report 26 (International Commission on Radiation Units and Measurements, Washington).

Ishihara, T., Iinuma, T. A., Tanaka, E. and Yashiro, S. (1969). "Plutonium lung monitor using a thin NaI(Tl) crystal of large area," Health Phys. **17,** 669.

IYENGAR, T. S., SADARANGANI, S. H., SOMASUNDARAM, S. AND VAZE, P. K. (1965). "A cold strip apparatus for sampling tritium in air," Health Phys. **11,** 313.

JACKSON, S. (1966). "Creatinine in urine as an index of urinary excretion rate," Health Phys. **12,** 843.

JACKSON, S. AND DOLPHIN, G. W. (1966). "The estimation of internal radiation dose from metabolic and urinary excretion data for a number of important radionuclides," Health Phys. **12,** 481.

JAQUISH, R. E. AND MOGHISSI, A. A. (1973). "Survey of analytical methods for environmental monitoring of krypton-85," page 169 in *Noble Gases,* Report No. CONF 730915, Stanley, R. E. and Moghissi, A. A., Eds. (National Environmental Research Center, Las Vegas, Nevada).

KATHREN, R. N. (1967). Private communication.

KATHREN, R. L., RISING, F. L. AND LARSON, H. V. (1971). "K-fluorescence x-rays: A multi-use tool for health physics," Health Phys. **21,** 285.

KERR, G. D. AND STRICKLER, T. D. (1966). "The application of solid state nuclear track detectors to the Hurst threshold detector system," Health Phys. **12,** 1141.

KIEFER, H., MAUSHART, R. AND MEJDAHL, V. (1969). "Radiation protection dosimetry," page 557 in *Radiation Dosimetry,* Vol. III, Attix, F. H. and Tochilin, E., Eds. (Academic Press, New York).

KROHN, B. J., CHAMBERS, W. B. AND STORM, E. (1969). *The Photon Energy Response of Several Commercial Ionization Chambers, Geiger Counters, Scintillators, and Thermoluminescent Detectors,* Report No. LA-4052 (Los Alamos Scientific Laboratory, Los Alamos, New Mexico).

KUSNETZ, H. L. (1956). "Radon-daughters in mine atmospheres: A field method for determining concentrations," Am. Ind. Hyg. Assoc. Q. **17,** 85.

LADU, M., PELLICCIONI, M. AND ROTONDI, E. (1965). "Flat response to neutrons between 20 keV and 14 MeV of a BF_3 counter in a spherical hollow moderator," Nucl. Instrum. Methods **32,** 173.

LANGMEAD, W. A. AND FARMER, F. T. (1971). "Film badge position for x-ray workers," Br. J. Radiol. **44,** 400.

LARSON, H. V. AND KEENE, A. R. (1965). "The Hanford emergency dosimetry system," page 673 in *Personnel Dosimetry for Radiation Accidents,* IAEA Publication No. STI/PUB/99 (International Atomic Energy Agency, Vienna).

LARSON, H. V., MEYERS, I. T. AND ROESCH, W. C. (1954). *A Wide Beam Fluorescent X-Ray Source,* USAEC Report No. HW-31781 (General Electric Company, Hanford Atomic Products Operation, Richland, Washington).

LAURER, G. R. AND EISENBUD, M. (1963). "Low-level *in vivo* measurement of iodine-131 in humans," Health Phys. **9,** 401.

LAURER, G. R. AND EISENBUD, M. (1968). "*In vivo* measurement of nuclides emitting soft penetrating radiations," page 189 in *Diagnosis and Treatment of Deposited Radionuclides,* Kornberg, H. A. and Norwood, W. D., Eds. (Excerpta Medica Foundation, New York).

LAWSON, R. C. AND WATT, D. E. (1964). "Neutron depth dose measurements in a tissue-equivalent phantom for an incident Pu-Be spectrum," Phys. Med. Biol. **9,** 487.

LEAKE, J. W. (1968). "Improved spherical dose equivalent neutron detector," Nucl. Instrum. Methods **63,** 329.

LEHMAN, R. L. (1970). "X-ray measurements around high-power klystrons," page 183 in *Electronic Product Radiation and the Health Physicist,* DHEW Publication No. BRH/DEP 70-26 (U.S. Public Health Service, Environmental Health Service, Bureau of Radiological Health, Rockville, Maryland).

LIEBERMAN, R. AND MOGHISSI, A. A. (1970). "Low-level counting by liquid scintillation. II: Applications of emulsions in tritium counting," Int. J. Appl. Radiat. Isot. **21,** 319.

LINDELL, B. (1968). "Occupational hazards in x-ray analytical work," Health Phys. **15,** 481.

LUCAS, H. F. (1957). "Improved low-level alpha-scintillation counter for radon," Rev. Sci. Instrum. **28,** 680.

MARINELLI, L. D., MILLER, C. E., MAY, H. A. AND ROSE, J. E. (1962). "Low-level gamma-ray scintillation spectrometry: experimental requirements and biomedical applications," page 81 in *Advances in Biological and Medical Physics,* Vol. 8, Tobias, C. A. and Lawrence, J. H., Eds. (Academic Press, New York).

MARTZ, D. E., HOLLEMAN, D. F., MCCURDY, D. E. AND SCHIAGER, K. J. (1969). "Analysis of atmospheric concentrations of RaA, RaB and RaC by alpha spectroscopy," Health Phys. **17,** 131.

MCCALL, R. C. (1960). "Low background shields," Health Phys. **2,** 304.

MCCALL, R. C. (1975). Private communication.

MCCONNON, D. (1970). *Use of Water as a Sampling Medium for Tritium Oxide,* Report No. BNWL-CC-547 (Battelle Pacific Northwest Laboratories, Richland, Washington).

MENEELY, G. R. AND LINE, S. M., Eds. (1965). *Radioactivity in Man. Second Symposium, Whole Body Counting and Effects of Internal Gamma Ray Emitting Radioisotopes,* (Charles C Thomas, Springfield, Illinois).

MOGHISSI, A. A., KELLEY, H. L., REGNIER, J. E. AND CARTER, M. W. (1969). "Low-level counting by liquid scintillation. I: Tritium measurements in homogeneous systems," Int. J. Appl. Radiat. Isot. **20,** 145.

MOORE, T. M., GUNDAKER, W. E. AND THOMAS, J. W., Eds. (1971). *Radiation Safety in X-Ray Diffraction and Spectroscopy,* DHEW Publication No. (FDA) 72-8009, BRH/DEP 72-3 (Bureau of Radiological Health, Rockville, Maryland).

MORROW, P. E., GIBB, F. R., DAVIES, H., MITOLA, J., WOOD, D., WRAIGHT, N. AND CAMPBELL, H. S. (1967). "The retention and fate of inhaled plutonium dioxide in dogs," Health Phys. **13,** 113.

MÜLLER, J., DAVID, A., REJSKOVA, M. AND BREZIKOVA, D. (1961). "Chronic occupational exposure to strontium-90 and radium-226," Lancet ii, 129.

MURTHY, K. B. S., MUTHUKRISHMAN, G. AND SUNTA, C. M. (1967). "A counter for fast neutron dosimetry," Indian J. Pure Appl. Phys. **5,** 26.

NACHTIGALL, D. AND BURGER, G. (1972). "Dose equivalent determinations in neutron fields by means of moderator techniques," page 385 in *Topics in Radiation Dosimetry,* Attix, F. H., Ed. (Academic Press, New York).

NASH, A. E. AND ATTIX, F. H. (1971). "Use of LiF-Teflon discs in laminated identification cards for personnel accident dosimetry," Health Phys. **21,** 435.

NAVERSTEN, Y., MCCALL, R. C. AND LIDEN, K. (1963). "Semi-portable whole-body counter for cesium 137 and other gamma-emitting isotopes," Acta Radiol., Ther. Phys. Biol. **1,** 190.

NBS (1958). National Bureau of Standards, *Safe Design and Use of Industrial Beta-Ray Sources,* American Standards Association Report No. 254.2, published as National Bureau of Standards Handbook No. 66 (U.S. Government Printing Office, Washington).

NCRP (1951a). National Committee on Radiation Protection and Measurements, *Control and Removal of Radioactive Contamination in Laboratories,* NCRP Report No. 8, published as National Bureau of Standards Handbook No. 48 (National Council on Radiation Protection and Measurements, Washington).

NCRP (1951b). National Committee on Radiation Protection and Measurements, *Recommendations for Waste Disposal of Phosphorus-32 and Iodine-131 for Medical Users,* NCRP Report No. 9, published as National Bureau of Standards Handbook No. 49 (National Council on Radiation Protection and Measurements, Washington).

NCRP (1953). National Committee on Radiation Protection and Measurements, *Recommendations for the Disposal of Carbon-14 Wastes,* NCRP Report No. 12, published as National Bureau of Standards Handbook No. 53 (National Council on Radiation Protection and Measurements, Washington).

NCRP (1954b). National Committee on Radiation Protection and Measurements, *Radioactive Waste Disposal in the Ocean,* NCRP Report No. 16, published as National Bureau of Standards Handbook No. 58 (National Council on Radiation Protection and Measurements, Washington).

NCRP (1959). National Committee on Radiation Protection and Measurements, *Maximum Permissible Body Burdens and Maximum Permissible Concentrations of Radionuclides in Air and in Water for Occupational Exposure,* NCRP Report No. 22, published as National Bureau of Standards Handbook No. 69 (National Council on Radiation Protection and Measurements, Washington).

NCRP (1960). National Committee on Radiation Protection and Measurements, *Measurement of Neutron Flux and Spectra for Physical and Biological Applications,* NCRP Report No. 23, published as National Bureau of Standards Handbook No. 72 (National Council on Radiation Protection and Measurements, Washington).

NCRP (1961a). National Committee on Radiation Protection and Measurements, *Measurement of Absorbed Dose of Neutrons and of Mixtures of Neutrons and Gamma Rays,* NCRP Report No. 25, published as National Bureau of Standards Handbook No. 75 (National Council on Radiation Protection and Measurements, Washington).

NCRP (1961b). National Committee on Radiation Protection and Measurements, *A Manual of Radioactivity Procedures,* NCRP Report No. 28, published as National Bureau of Standards Handbook No. 80 (National Council on Radiation Protection and Measurements, Washington).

NCRP (1963). National Committee on Radiation Protection and Measurements, *Addendum 1 to National Bureau of Standards Handbook 69,*

Maximum Permissible Body Burdens and Maximum Permissible Concentrations of Radionuclides in Air and in Water for Occupational Exposure (National Council on Radiation Protection and Measurements, Washington).

NCRP (1964). National Committee on Radiation Protection and Measurements, *Safe Handling of Radioactive Materials,* NCRP Report No. 30, published as National Bureau of Standards Handbook No. 92 (National Council on Radiation Protection and Measurements, Washington).

NCRP (1968a). National Council on Radiation Protection and Measurements, *Medical X-Ray and Gamma-Ray Protection for Energies up to 10 MeV—Equipment Design and Use,* NCRP Report No. 33 (National Council on Radiation Protection and Measurements, Washington).

NCRP (1968b). National Council on Radiation Protection and Measurements, *X-Ray Protection Standards for Home Television Receivers,* Interim Statement (National Council on Radiation Protection and Measurements, Washington).

NCRP (1970). National Council on Radiation Protection and Measurements, *Dental X-Ray Protection,* NCRP Report No. 35 (National Council on Radiation Protection and Measurements, Washington).

NCRP (1971a). National Council on Radiation Protection and Measurements, *Protection Against Neutron Radiation,* NCRP Report No. 38 (National Council on Radiation Protection and Measurements, Washington).

NCRP (1971b). National Council on Radiation Protection and Measurements, *Basic Radiation Protection Criteria,* NCRP Report No. 39 (National Council on Radiation Protection and Measurements, Washington).

NCRP (1972). National Council on Radiation Protection and Measurements, *Protection Against Radiation from Brachytherapy Sources,* NCRP Report No. 40 (National Council on Radiation Protection and Measurements, Washington).

NCRP (1976a). National Council on Radiation Protection and Measurements, *Tritium Measurement Techniques,* NCRP Report No. 47 (National Council on Radiation Protection and Measurements, Washington).

NCRP (1976b). National Council on Radiation Protection and Measurements, *Structural Shielding Design and Evaluation for Medical Use of X-Rays and Gamma Rays of Energies Up to 10 MeV,* NCRP Report No. 49 (National Council on Radiation Protection and Measurements, Washington).

NCRP (1976c). National Council on Radiation Protection and Measurements, *Environmental Radiation Measurements*, NCRP Report No. 50. (National Council on Radiation Protection and Measurements, Washington).

NCRP (1977). National Council on Radiation Protection and Measurements, *Radiation Protection Design Guidelines for 0.1–100 MeV Particle Accelerator Facilities,* NCRP Report No. 51 (National Council on Radiation Protection and Measurements, Washington).

NCRP (1978). National Council on Radiation Protection and Measurements. *Manual of Radioactivity Measurements Procedures*, NCRP Report No. 58. (National Council on Radiation Protection and Measurements, Washington).

NELLIS, D. O., HUDSPETH, E. L., MORGAN, I. L., BUCHANAN, P. S. AND BOGGS, R. F. (1967). *Tritium Contamination in Particle Accelerator Operation,* PHS Publication No. 999-RH-29 (U.S. Department of Health, Education, and Welfare, Washington).

OSBORNE, R. V. AND COVEART, A. S. (1973). "Portable monitor for tritium in air," Nucl. Instrum. Methods **106,** 181.

OSLOOND, J. H., ECHO, J. B., POLZER, W. L. AND JOHNSON, B. D. (1973). "A tritium air sampling method for environmental and nuclear plant monitoring," page 531 in *Tritium,* Moghissi, A. A. and Carter, M. W., Eds. (Messenger Graphics, Phoenix, Arizona).

PALMER, H. E. (1966). "Determination of ^{32}P *in vivo,*" Health Phys. **12,** 605.

PALMER, H. E. AND ROESCH, W. C. (1965). "A shadow-shield whole-body counter," Health Phys. **11,** 1213.

PARKER, D. AND ANDERSON, J. I. (1967). "A portable whole-body counter," Health Phys. **13,** 141.

PARSONS, D. F., PHILLIPS, V. A. AND LALLY, J. S. (1974). "The radiological significance of x-ray leakage from electron microscopes," Health Phys. **26,** 439.

PATTERSON, H. W. AND THOMAS, R. H. (1973). *Accelerator Health Physics* (Academic Press, New York).

PETERSEN, D. F., MITCHELL, V. E. AND LANGHAM, W. H. (1961). "Estimation of fast neutron doses in man by $S^{32}(n,p)P^{32}$ reaction in body hair," Health Phys. **6,** 1.

PIESCH, E. (1969). "Überprufüng der Y-Diskriminierung von rem-countern," Direct Inform. 6/69.

PIESCH, E. (1972). "Developments in radiophotoluminescence dosimetry," Chapter 8 in *Topics in Radiation Dosimetry,* Attix, F. H., Ed. (Academic Press, New York).

POCHIN, E. E. (1964). "Formulation of relationships between the radiation exposure of tissues and the excretion rate of nuclides," page 3 in *Assessment of Radioactivity in Man,* Vol. 1, IAEA Publication No. STI/PUB/84 (International Atomic Energy Agency, Vienna).

PROPERZIO, W. S. (1970). "An inexpensive rf shield for use with radiation survey instruments," Health Phys. **19,** 442.

RAMM, W. J. (1966). "Scintillation detectors," page 142 in *Radiation Dosimetry,* Vol. II, Attix, F. H. and Roesch, W. C., Eds. (Academic Press, New York).

RAMSDEN, D. (1969). "The measurement of plutonium-239 *in vivo,*" Health Phys. **16,** 145.

RAMSDEN, D. (1976). "Assessment of plutonium in lung for both chronic and acute exposure conditions," page 139 in *Diagnosis and Treatment of Incorporated Radionuclides,* IAEA Publication No. STI/PUB/411 (International Atomic Energy Agency, Vienna).

REINHARDT, P. W. AND DAVIS, F. J. (1958). "Improvements in the threshold detector method of fast neutron dosimetry," Health Phys. **1,** 169.

REINIG, W. C. AND ALBANESIUS, E. L. (1962). *Control of Tritium Health Hazards at Savannah River,* Report No. DPSPU 62-30-5 (DuPont de Nemours & Co., Savannah River Laboratory, Aiken, South Carolina), also

Am. Ind. Hyg. Assoc. J. **24,** 276.

REINIG, W. C., WRIGHT, C. N. AND HOY, J. E. (1965). "A pocket dosimeter for criticality accidents," page 307 in *Personnel Dosimetry for Radiation Accidents,* IAEA Publication No. STI/PUB/99 (International Atomic Energy Agency, Vienna).

ROBERTSON, M. K. AND RANDLE, M. W. (1974). "Hazards from the industrial use of radioactive static eliminators," Health Phys. **26,** 245.

ROESCH, W. C. AND BAUM, J. W. (1958). "Detection of plutonium in wounds," page 142 in *Proceedings of the 2nd International Conference on the Peaceful Uses of Atomic Energy,* Vol. 23, Geneva (United Nations, New York).

ROESCH, W. C. AND DONALDSON, E. E. (1958). "Portable instruments for beta-ray dosimetry," page 172 in *Proceedings of the International Conference on Peaceful Uses of Atomic Energy,* Vol. 14, Geneva (Columbia University Press, New York).

ROSSI, H. H. AND ROSENZWEIG, W. (1955). "A device for the measurement of dose as a function of specific ionization," Radiology **64,** 404.

ROTONDI, E. (1966). *A Neutron Fluence Meter with Uniform Angular Response,* Report No. APXNR-1849, NRC-9102 (National Research Council of Canada, Ottawa).

ROTONDI, E. AND GEIGER, K. W. (1968). "A neutron dosemeter with spherical moderator containing absorbers and a BF_3 counter," page 141 in *Proceedings of the First International Congress of Radiation Protection,* Vol. I, Snyder, W. S., Abee, H. H., Burton, L. K., Maushart, R., Benco, A., Duhamel, F. and Wheatley, B. M., Eds. (Pergamon Press, New York).

RUNDO, J. (1962). "Some calibration problems of whole-body gamma-ray spectrometers," page 121 in *Whole-Body Counting,* IAEA Publication No. STI/PUB/47 (International Atomic Energy Agency, Vienna).

SANDERS, S. M. (1960). "Power functions relating excretion to body burden," Health Phys. **2,** 295.

SEHMEL, G. A. (1967a). "Estimation of airstream concentrations of particulates from sub-isokinetically obtained filter samples," Am. Ind. Hyg. Assoc. J. **28,** 243.

SEHMEL, G. A. (1967b). "Validity of air samples as affected by anisokinetic sampling and deposition within sampling line," page 727 in *Assessment of Airborne Radioactivity,* IAEA Publication No. STI/PUB/159 (International Atomic Energy Agency, Vienna).

SILL, C. W., ANDERSON, J. I. AND PERCIVAL, D. R. (1964). "Comparison of excretion analysis with whole-body counting for assessment of internal radioactive contaminants," page 217 in *Assessment of Radioactivity in Man,* Vol. I, IAEA Publication No. STI/PUB/84 (International Atomic Energy Agency, Vienna).

SINCLAIR, W. K. (1950). "Comparison of Geiger-counter and ion-chamber methods of measuring gamma radiation," Nucleonics **7,** No. 6, 21.

SMITH, A. R. (1961). "A cobalt neutron-flux integrator," Health Phys. **7,** 40.

SMITH, J. W. (1962). "Sodium activation by fast neutrons in man phantoms," Phys. Med. Biol. **7,** 341.

STOMS, R. AND KUERZE, E. (1961). "Instrument for measurement of x-ray emission from television sets," Department of Health, Education, and Welfare Report No. TSB-1 (National Technical Information Service, Springfield, Virginia).

STORM, E., LIER, D. W. AND ISRAEL, H. I. (1974). "Photon sources for instrument calibration," Health Phys. **26,** 179.

STRAUB, C. P. (1964). *Low-Level Radioactive Wastes: Their Handling, Treatment and Disposal,* Chapter 3 (U.S. Government Printing Office, Washington).

SWINTH, K. L. AND GRIFFIN, B. I. (1970). "A developmental scintillation counter for detection of plutonium *in vivo,*" Health Phys. **19,** 543.

THOMAS, J. W. AND LECLARE, P. C. (1970). "A study of the two-filter method for radon-222," Health Phys. **18,** 133.

THOMPSON, J. J. AND ZIEMER, P. L. (1972). "Energy response of lithium borate thermoluminescent dosimeters," Health Phys. **22,** 399.

TOMLINSON, F. K. (1976). "Plutonium-238 lung counting with Phoswich detectors," page 237 in *Diagnosis and Treatment of Incorporated Radionuclides,* IAEA Publication No. STI/PUB/411 (International Atomic Energy Agency, Vienna).

TROTT, N. G., PARNELL, C. J., HODT, H. J. AND ENTWHISTLE, R. F. (1963). "Studies in the design and applications of a clinical low-background counting room," Br. J. Radiol. **36,** 592.

TSIVOGLOU, E. C., AYER, H. E. AND HOLADAY, D. A. (1953). "Occurrence of nonequilibrium atmospheric mixtures of radon and its daughters," Nucleonics **11,** No. 9, 40.

Union Carbide (1958). *Accidental Radiation Excursion at the Y-12 Plant, June 16, 1958,* Report No. Y-1234 (Union Carbide Nuclear Co., Oak Ridge, Tennessee).

VENNART, J., MAYCOCK, G., GODFREY, B. E. AND DAVIES, B. L. (1964). "Measurements of radium in radium luminizers," page 277 in *Assessment of Radioactivity in Man,* Vol. II, IAEA Publication No. STI/PUB/84 (International Atomic Energy Agency, Vienna).

VOELZ, G., UMBARGER, J., MCINROY, J. AND HEALY, J. (1976). "Considerations in the assessment of plutonium deposition in man," page 163 in *Diagnosis and Treatment of Incorporated Radionuclides,* IAEA Publication No. STI/PUB/411 (International Atomic Energy Agency, Vienna).

WAGNER, E. B. AND HURST, G. S. (1958). "Advances in the standard proportional counter method of fast neutron dosimetry," Rev. Sci. Instrum. **29,** 153.

WATSON, E. E., CLOUTIER, R. J. AND BERGER, J. D. (1969). "Tritium concentrations in neutron generator pump oil," Health Phys. **17,** 739.

WEBB, G. A. M., DAUCH, J. E. AND BODIN, G. (1972). "Operational evaluation of a new high sensitivity thermoluminescent dosimeter," Health Phys. **23,** 89.

WEBSTER, E. W. (1969). "Patient and personnel dose during diagnostic x-ray examinations," page 96 in *Medical Radiation Information for Litigation,* DHEW Publication No. DMRE 69-3 (Bureau of Radiological Health, Rockville, Maryland).

WELLMAN, H. N., KEREIAKES, J. G., YEAGER, T. B., KARCHES, G. J. AND SAENGER, E. L. (1967). "A sensitive technique for measuring thyroidal uptake of iodine-131," J. Nucl. Med. **8,** 86.

WHO (1966). World Health Organization, *Methods of Radiochemical Analysis: Report of a Scientific Meeting* (World Health Organization, Geneva).

WILENZICK, R. M., ALMOND, P. R., OLIVER, G. D. AND DE ALMEIDA, C. E. (1973). "Measurement of fast neutrons produced by high-energy x-ray beams of medical electron accelerators," Phys. Med. Biol. **18,** 396.

WILSON, R. H. (1962). *A Method for Immediate Detection of High-Level Neutron Exposure by Measurement of Sodium-24 in Humans,* Document No. HW-73891 (Revised) (General Electric Company, Hanford Atomic Products Operation, Richland, Washington).

WOLD, G. J., SCHEELE, R. V. AND AGARWAL, S. K. (1971). "Evaluation of physician exposure during cardiac catheterization," Radiology **99,** 188.

WOLLAN, R. O., STAIGER, J. W. AND BOGE, R. J. (1971). "Disposal of low-level radioactive wastes at a large university incinerator," Am. Ind. Hyg. Assoc. J. **32,** 625.

WOOD, V. A. (1968). *A Collection of Radium Leak Test Articles,* Technical Report No. MORP 68-1 (Bureau of Radiological Health, Rockville, Maryland).

WYCKOFF, H. O. (1950). "Calibration of instruments for measuring ionizing radiations," page 950 in *The Encyclopedia of Instrumentation for Industrial Hygiene* (University of Michigan, Ann Arbor, Michigan).

YAFFE, L. (1962). "Preparation of thin films, sources, and targets," Annu. Rev. Nucl. Sci. **12,** 153.

YOKOTA, R. AND NAKAJIMA, S. (1965). "Improved fluoroglass dosimeter as personnel monitoring dosimeter and microdosimeter," Health Phys. **11,** 241.

ZELAC, R. E. (1968). "Track fading in neutron personal monitoring film under typical use conditions," Health Phys. **15,** 545.

The NCRP

The National Council on Radiation Protection and Measurements is a nonprofit corporation chartered by Congress in 1964 to:

1. Collect, analyze, develop, and disseminate in the public interest information and recommendations about (a) protection against radiation and (b) radiation measurements, quantities, and units, particularly those concerned with radiation protection;
2. Provide a means by which organizations concerned with the scientific and related aspects of radiation protection and of radiation quantities, units, and measurements may cooperate for effective utilization of their combined resources, and to stimulate the work of such organizations;
3. Develop basic concepts about radiation quantities, units, and measurements, about the application of these concepts, and about radiation protection;
4. Cooperate with the International Commission on Radiological Protection, the International Commission on Radiation Units and Measurements, and other national and international organizations, governmental and private, concerned with radiation quantities, units, and measurements and with radiation protection.

The Council is the successor to the unincorporated association of scientists known as the National Committee on Radiation Protection and Measurements and was formed to carry on the work begun by the Committee.

The Council is made up of the members and the participants who serve on the fifty-eight Scientific Committees of the Council. The Scientific Committees, composed of experts having detailed knowledge and competence in the particular area of the Committee's interest, draft proposed recommendations. These are then submitted to the full membership of the Council for careful review and approval before being published.

The following comprise the current officers and members of the Council:

Officers

President	Warren K. Sinclair
Vice President	Hymer L. Friedell
Secretary and Treasurer	W. Roger Ney
Assistant Secretary	Eugene R. Fidell
Assistant Treasurer	Harold O. Wyckoff

Members

Seymour Abrahamson
S. James Adelstein
Roy E. Albert
Frank H. Attix
John A. Auxier
William J. Bair
Victor P. Bond
Harold S. Boyne
Robert L. Brent
A. Bertrand Brill
Reynold F. Brown
William W. Burr, Jr.
Melvin W. Carter
George W. Casarett
Randall S. Caswell
Arthur B. Chilton
Stephen F. Cleary
Cyril L. Comar
Patricia W. Durbin
Merril Eisenbud
Thomas S. Ely
Benjamin G. Ferris
Asher J. Finkel
Donald C. Fleckenstein
Richard F. Foster
Hymer L. Friedell
Arthur H. Gladstein
Barry B. Goldberg
Robert A. Goepp
Marvin Goldman
Robert O. Gorson
Douglas Grahn
Arthur W. Guy
Ellis M. Hall
John H. Harley
John W. Healy
Louis H. Hempelmann, Jr.
John M. Heslep
George B. Hutchison
Marylou Ingram
Seymour Jablon
Jacob Kastner
Thomas A. Lincoln
Edward B. Lewis
Charles W. Mays
Roger O. McClellan
James McLaughlin
Charles B. Meinhold
Mortimer M. Mendelsohn
Dade W. Moeller
Paul E. Morrow
Robert D. Moseley, Jr.
James V. Neel
Robert J. Nelsen
Frank Parker
Andrew K. Poznanski
William C. Reinig
Chester R. Richmond
Harald H. Rossi
Robert E. Rowland
Eugene L. Saenger
Harry F. Schulte
Raymond Seltser
Warren K. Sinclair
J. Newell Stannard
Chauncey starr
John B. Storer
Roy C. Thompson
James E. Turner
John C. Villforth
George L. Voelz
Niel Wald
Edward W. Webster
George M. Wilkening
McDonald E. Wrenn

Honorary Members

Currently, the following Scientific Committees are actively engaged in formulating recommendations:

- SC-1: Basic Radiation Protection Criteria
- SC-3: Medical X- and Gamma-Ray Protection Up to 10 MeV (Equipment Design and Use)
- SC-11: Incineration of Radioactive Waste
- SC-16: X-Ray Protection in Dental Offices
- SC-18: Standards and Measurements of Radioactivity for Radiological Use
- SC-24: Radionuclides and Labeled Organic Compounds Incorporated in Genetic Material
- SC-25: Radiation Protection in the Use of Small Neutron Generators
- SC-26: High Energy X-Ray Dosimetry
- SC-30: Physical and Biological Properties of Radionuclides
- SC-32: Administered Radioactivity
- SC-33: Dose Calculations
- SC-34: Maximum Permissible Concentrations for Occupational and Non-Occupational Exposures
- SC-37: Procedures for the Management of Contaminated Persons
- SC-38: Waste Disposal
- SC-39: Microwaves
- SC-40: Biological Aspects of Radiation Protection Criteria
- SC-41: Radiation Resulting from Nuclear Power Generation
- SC-42: Industrial Applications of X Rays and Sealed Sources
- SC-44: Radiation Associated with Medical Examinations
- SC-45: Radiation Received by Radiation Employees
- SC-46: Operational Radiation Safety
- SC-47: Instrumentation for the Determination of Dose Equivalent
- SC-48: Apportionment of Radiation Exposure
- SC-50: Surface Contamination
- SC-51: Radiation Protection in Pediatric Radiology and Nuclear Medicine Applied to Children
- SC-52: Conceptual Basis of Calculations of Dose Distributions
- SC-53: Biological Effects and Exposure Criteria for Radiofrequency Electromagnetic Radiation
- SC-54: Bioassay for Assessment of Control of Intake of Radionuclides
- SC-55: Experimental Verification of Internal Dosimetry Calculations
- SC-56: Mammography
- SC-57: Internal Emitter standards
- SC-58: Radioactivity in water

In recognition of its responsibility to facilitate and stimulate cooperation among organizations concerned with the scientific and related aspects of radiation protection and measurement, the Council has created a category of NCRP Collaborating Organizations. Organizations or groups of organizations that are national or international in scope and are concerned with scientific problems involving radiation quantities, units, measurements and effects, or radiation protection may be admitted to collaborating status by the Council. The present Collaborating Organizations with which the NCRP maintains liaison are as follows:

American Academy of Dermatology
American Association of Physicists in Medicine
American College of Radiology
American Dental Association
American Industrial Hygiene Association
American Insurance Association
American Medical Association
American Nuclear Society
American Occupational Medical Association
American Podiatry Association
American Public Health Association
American Radium Society
American Roentgen Ray Society
American Society of Radiologic Technologists
Association of University Radiologists
Atomic Industrial Forum
College of American Pathologists
Defense Civil Preparedness Agency
Genetics Society of America
Health Physics Society
National Bureau of Standards
National Electrical Manufacturers Association
Radiation Research Society
Radiological Society of North America
Society of Nuclear Medicine
United States Air Force
United States Army
United States Department of Energy
United States Environmental Protection Agency
United States Navy
United States Nuclear Regulatory Commission
United States Public Health Service

The NCRP has found its relationships with these organizations to be extremely valuable to continued progress in its program.

The Council's activities are made possible by the voluntary contribution of the time and effort of its members and participants and the generous support of the following organizations:

Alfred P. Sloan Foundation
Alliance of American Insurers
American Academy of Dental Radiology
American Academy of Dermatology
American Association of Physicists in Medicine
American College of Radiology
American College of Radiology Foundation
American Dental Association
American Industrial Hygiene Association
American Insurance Association
American Medical Association
American Nuclear Society
American Occupational Medical Association
American Osteopathic College of Radiology
American Podiatry Association
American Public Health Association
American Radium Society
American Roentgen Ray Society
American Society of Radiologic Technologists
American Veterinary Medical Association
American Veterinary Radiology Society
Association of University Radiologists
Atomic Industrial Forum
Battelle Memorial Institute
College of American Pathologists
Defense Civil Preparedness Agency
Edward Mallinckrodt, Jr. Foundation
Electric Power Research Institute
Genetics Society of America
Health Physics Society
James Picker Foundation
National Association of Photographic Manufacturers
National Bureau of Standards
National Electrical Manufacturers Association
Radiation Research Society
Radiological Society of North America
Society of Nuclear Medicine
United States Department of Energy
United States Environmental Protection Agency
United States Navy
United States Nuclear Regulatory Commission
United States Public Health Service

To all of these organizations the Council expresses its profound appreciation for their support.

Initial funds for publication of NCRP reports were provided by a grant from the James Picker Foundation and for this the Council wishes to express its deep appreciation.

The NCRP seeks to promulgate information and recommendations based on leading scientific judgment on matters of radiation protection and measurement and to foster cooperation among organizations concerned with these matters. These efforts are intended to serve the public interest and the Council welcomes comments and suggestions on its reports or activities from those interested in its work.

NCRP Publications

NCRP publications are distributed by the NCRP Publications' office. Information on prices and how to order may be obtained by directing an inquiry to:

NCRP Publications
P.O. Box 30175
Washington, D.C. 20014

The extant publications are listed below.

Lauriston S. Taylor Lectures

No.	Title and Author
1	*The Squares of the Natural Numbers in Radiation Protection* by Herbert M. Parker
2	*Why be Quantitative about Radiation Risk Estimates* by Sir Edward E. Pochin

NCRP Reports

No.	Title
8	*Control and Removal of Radioactive Contamination in Laboratories* (1951)
9	*Recommendations for Waste Disposal of Phosphorus-32 and Iodine-131 for Medical Users* (1951)
12	*Recommendations for the Disposal of Carbon-14 Wastes* (1953)
16	*Radioactive Waste Disposal in the Ocean* (1954)
22	*Maximum Permissible Body Burdens and Maximum Permissible Concentrations of Radionuclides in Air and in Water for Occupational Exposure* (1959) [Includes Addendum 1 issued in August 1963]
23	*Measurement of Neutron Flux and Spectra for Physical and Biological Applications* (1960)

25	*Measurement of Absorbed Dose of Neutrons and of Mixtures of Neutrons and Gamma Rays* (1961)
27	*Stopping Powers for Use with Cavity Chambers* (1961)
28	*A Manual of Radioactivity Procedures* (1961)
30	*Safe Handling of Radioactive Materials* (1964)
32	*Radiation Protection in Educational Institutions* (1966)
33	*Medical X-Ray and Gamma-Ray Protection for Energies Up to 10 MeV—Equipment Design and Use* (1968)
35	*Dental X-Ray Protection* (1970)
36	*Radiation Protection in Veterinary Medicine* (1970)
37	*Precautions in the Management of Patients Who Have Received Therapeutic Amounts of Radionuclides* (1970)
38	*Protection against Neutron Radiation* (1971)
39	*Basic Radiation Protection Criteria* (1971)
40	*Protection Against Radiation From Brachytherapy Sources* (1972)
41	*Specification of Gamma-Ray Brachytherapy Sources* (1974)
42	*Radiological Factors Affecting Decision-Making in a Nuclear Attack* (1974)
43	*Review of the Current State of Radiation Protection Philosophy* (1975)
44	*Krypton-85 in the Atmosphere—Accumulation, Biological Significance, and Control Technology* (1975)
45	*Natural Background Radiation in the United States* (1975)
46	*Alpha-Emitting Particles in Lungs* (1975)
47	*Tritium Measurement Techniques* (1976)
48	*Radiation Protection for Medical and Allied Health Personnel* (1976)
49	*Structural Shielding Design and Evaluation for Medical Use of X Rays and Gamma Rays of Energies Up to 10 MeV* (1976)
50	*Environmental Radiation Measurements* (1976)
51	*Radiation Protection Design Guidelines for 0.1-100 MeV Particle Accelerator Facilities* (1977)
52	*Cesium-137 From the Environment to Man: Metabolism and Dose* (1977)
53	*Review of NCRP Radiation Dose Limit for Embryo and Fetus in Occupationally-Exposed Women* (1977)
54	*Medical Radiation Exposure of Pregnant and Potentially Pregnant Women* (1977)

55 *Protection of the Thyroid Gland in the Event of Releases of Radioiodine* (1977)

56 *Radiation Exposure From Consumer Products and Miscellaneous Sources* (1977)

57 *Instrumentation and Monitoring Methods for Radiation Protection* (1978)

Binders for NCRP Reports are available. Two sizes make it possible to collect into small binders the "old series" of reports (NCRP Reports Nos. 8-31) and into large binders the more recent publications (NCRP Reports Nos. 32-57). Each binder will accommodate from five to seven reports. The binders carry the identification "NCRP Reports" and come with label holders that permit the user to attach labels showing the reports contained in each binder.

The following bound sets of NCRP Reports are also available:

Volume I. NCRP Reports Nos. 8, 9, 12, 16, 22
Volume II. NCRP Reports Nos. 23, 25, 27, 28, 30
Volume III. NCRP Reports Nos. 32, 33, 35, 36, 37
Volume IV. NCRP Reports Nos. 38, 39, 40, 41
Volume V. NCRP Reports Nos. 42, 43, 44, 45, 46
Volume VI. NCRP Reports Nos. 47, 48, 49, 50, 51

(Titles of the individual reports contained in each volume are given above.)

The following NCRP reports are now superseded and/or out of print:

NCRP Report No.	Title
1	*X-Ray Protection* (1931). [Superseded by NCRP Report No. 3]
2	*Radium Protection* (1934). [Superseded by NCRP Report No. 4]
3	*X-Ray Protection* (1936). [Superseded by NCRP Report No. 6]
4	*Radium Protection* (1938). [Superseded by NCRP Report No. 13]
5	*Safe Handling of Radioactive Luminous Compounds* (1941). [Out of print]
6	*Medical X-Ray Protection up to Two Million Volts* (1949). [Superseded by NCRP Report No. 18]
7	*Safe Handling of Radioactive Isotopes* (1949). [Superseded by NCRP Report No. 30]
10	*Radiological Monitoring Methods and Instruments* (1952). [Superseded by NCRP Report No. 57]
11	*Maximum Permissible Amounts of Radioisotopes in the Human Body and Maximum Permissible Concentrations in*

Air and Water (1953). [Superseded by NCRP Report No. 22]

13 *Protection Against Radiations from Radium, Cobalt-60 and Cesium-137* (1954). [Superseded by NCRP Report No. 24]

14 *Protection Against Betatron—Synchrotron Radiations Up to 100 Million Electron Volts* (1954). [Superseded by NCRP Report No. 51].

15 *Safe Handling of Cadavers Containing Radioactive Isotopes* (1953). [Superseded by NCRP Report No. 21]

17 *Permissible Dose from External Sources of Ionizing Radiation* (1954) including *Maximum Permissible Exposure to Man, Addendum to National Bureau of Standards Handbook 59* (1958). [Superseded by NCRP Report No. 39]

18 *X-Ray Protection* (1955). [Superseded by NCRP Report No. 26]

19 *Regulation of Radiation Exposure by Legislative Means* (1955). [Out of print]

20 *Protection Against Neutron Radiation Up to 30 Million Electron Volts* (1957). [Superseded by NCRP Report No. 38]

21 *Safe Handling of Bodies Containing Radioactive Isotopes* (1958) [Superseded by NCRP Report No. 37]

24 *Protection Against Radiations from Sealed Gamma Sources* (1960). [Superseded by NCRP Reports Nos. 33, 34, and 40]

26 *Medical X-Ray Protection Up to Three Million Volts* (1961). [Superseded by NCRP Reports Nos. 33, 34, 35, and 36]

29 *Exposure to Radiation in an Emergency* (1962). [Superseded by NCRP Report No. 42]

31 *Shielding for High Energy Electron Accelerator Installations* (1964). [Superseded by NCRP Report No. 51]

34 *Medical X-Ray and Gamma-Ray Protection for Energies Up to 10 MeV—Structural Shielding Design and Evaluation* (1970). [Superseded by NCRP Report No. 49]

The following statements of the NCRP were published outside of the NCRP Report series:

"Blood Counts, Statement of the National Committee on Radiation Protection," Radiology 63, 428 (1954)

"Statements on Maximum Permissible Dose from Television Receivers and Maximum Permissible Dose to the Skin of the Whole Body," Am. J. Roentgenol., Radium Ther. and Nucl. Med. 84, 152 (1960) and Radiology 75, 122 (1960)

X-Ray Protection Standards for Home Television Receivers, Interim Statement of the National Council on Radiation Protection and Measurements (National Council on Radiation Protection and Measurements, Washington, 1968)

Specification of Units of Natural Uranium and Natural Thorium (National Council on Radiation Protection and Measurements, Washington, 1973)

Copies of the statements published in journals may be consulted in libraries. A limited number of copies of the last two statements listed above are available for distribution by NCRP Publications.

Index